Deep Learning with Applications Using Python

Chatbots and Face, Object, and Speech Recognition With TensorFlow and Keras

Navin Kumar Manaswi

Apress®

Deep Learning with Applications Using Python

Navin Kumar Manaswi
Bangalore, Karnataka, India

ISBN-13 (pbk): 978-1-4842-3515-7   ISBN-13 (electronic): 978-1-4842-3516-4
https://doi.org/10.1007/978-1-4842-3516-4

Library of Congress Control Number: 2018938097

Managing Director, Apress Media LLC: Welmoed Spahr
Acquisitions Editor: Celestin Suresh John
Development Editor: Matthew Moodie
Coordinating Editor: Divya Modi

Cover designed by eStudioCalamar

Cover image designed by Freepik (www.freepik.com)

Distributed to the book trade worldwide by Springer Science+Business Media New York, 233 Spring Street, 6th Floor, New York, NY 10013. Phone 1-800-SPRINGER, fax (201) 348-4505, e-mail orders-ny@springer-sbm.com, or visit www.springeronline.com. Apress Media, LLC is a California LLC and the sole member (owner) is Springer Science + Business Media Finance Inc (SSBM Finance Inc). SSBM Finance Inc is a **Delaware** corporation.

For information on translations, please e-mail rights@apress.com, or visit www.apress.com/rights-permissions.

Apress titles may be purchased in bulk for academic, corporate, or promotional use. eBook versions and licenses are also available for most titles. For more information, reference our Print and eBook Bulk Sales web page at www.apress.com/bulk-sales.

Any source code or other supplementary material referenced by the author in this book is available to readers on GitHub via the book's product page, located at www.apress.com/9781484235157. For more detailed information, please visit www.apress.com/source-code.

Printed on acid-free paper

Table of Contents

Foreword

Deep Learning has come a really long way. From the birth of the idea to understand human mind and the concept of associationism — how we perceive things and how relationships of objects and views influence our thinking and doing, to the modelling of associationism which started in the 1870s when Alexander Bain introduced the first concert of Artificial Neural Networks by grouping the neurons.

Fast forward it to today 2018 and we see how Deep Learning has dramatically improved and is in all forms of life — from object detection, speech recognition, machine translation, autonomous vehicles, face detection and the use of face detection from mundane tasks such as unlocking your iPhoneX to doing more profound tasks such as crime detection and prevention.

Convolutional Neural Networks and Recurrent Neural Networks are shining brightly as they continue to help solve the world problems in literally all industry areas such as Automotive & Transportation, Healthcare & Medicine, Retail to name a few. Great progress is being made in these areas and just metrics like these say enough about the palpability of the deep learning industry:

- Number of Computer Science academic papers have soared to almost 10x since 1996

- VCs are investing 6x more in AI startups since 2000

- There are 14x more active AI startups since 2000

- AI related jobs market is hiring 5x more since 2013 and Deep Learning is the most sought after skill in 2018

– 84% of enterprises believe investing in AI will give them a great competitive edge

And finally,

– the error rate of image classification has dropped from 28% in 2012 to 2.5% in 2017 and it is going down all the time!

Still the research community is not satisfied. We are pushing boundaries and I am moving ahead with my peers to develop models around the bright and shiny Capsule Networks and give Deep Learning a huge edge. I am not the only one in this battle. It is with great pleasure I put this foreword for Navin, a respected professional in the Deep Learning community I have come to know so well.

His book is coming just at the right moment. The industry as well as learners are in need of practical means to strengthen their knowledge in Deep Learning and apply in their job.

I am convinced that Navin's book will give the learners what they need. TensorFlow is increasingly becoming the market leader and Keras too is being adopted by professionals to solve difficult problems in computer vision and NLP (Natural Language Processing). There is no single company on this planet who isn't investing in these two application areas.

I look forward to this book being published and will be the first in line to get it. And my advice to you is: you should too!

About the Author

Navin Kumar Manaswi has been developing AI solutions with the use of cutting-edge technologies and sciences related to artificial intelligence for many years. Having worked for consulting companies in Malaysia, Singapore, and the Dubai Smart City project, as well as his own company, he has developed a rare mix of skills for delivering end-to-end artificial intelligence solutions, including video intelligence, document intelligence, and human-like chatbots. Currently, he solves B2B problems in the verticals of healthcare, enterprise applications, industrial IoT, and retail at Symphony AI Incubator as a deep learning AI architect. With this book, he wants to democratize the cognitive computing and services for everyone, especially developers, data scientists, software engineers, database engineers, data analysts, and C-level managers.

About the Technical Reviewer

Sundar Rajan Raman has more than 14 years of full stack IT experience in machine learning, deep learning, and natural language processing. He has six years of big data development and architect experience, including working with Hadoop and its ecosystems as well as other NoSQL technologies such as MongoDB and Cassandra. In fact, he has been the technical reviewer of several books on these topics.

He is also interested in strategizing using Design Thinking principles and coaching and mentoring people.

CHAPTER 1

Basics of TensorFlow

This chapter covers the basics of TensorFlow, the deep learning framework. Deep learning does a wonderful job in pattern recognition, especially in the context of images, sound, speech, language, and time-series data. With the help of deep learning, you can classify, predict, cluster, and extract features. Fortunately, in November 2015, Google released TensorFlow, which has been used in most of Google's products such as Google Search, spam detection, speech recognition, Google Assistant, Google Now, and Google Photos. Explaining the basic components of TensorFlow is the aim of this chapter.

TensorFlow has a unique ability to perform partial subgraph computation so as to allow distributed training with the help of partitioning the neural networks. In other words, TensorFlow allows model parallelism and data parallelism. TensorFlow provides multiple APIs. The lowest level API—TensorFlow Core—provides you with complete programming control.

Note the following important points regarding TensorFlow:

- Its graph is a description of computations.

- Its graph has nodes that are operations.

- It executes computations in a given context of a session.

- A graph must be launched in a session for any computation.

© Navin Kumar Manaswi 2018
N. K. Manaswi, *Deep Learning with Applications Using Python*,
https://doi.org/10.1007/978-1-4842-3516-4_1

- A session places the graph operations onto devices such as the CPU and GPU.

- A session provides methods to execute the graph operations.

For installation, please go to https://www.tensorflow.org/install/. I will discuss the following topics:

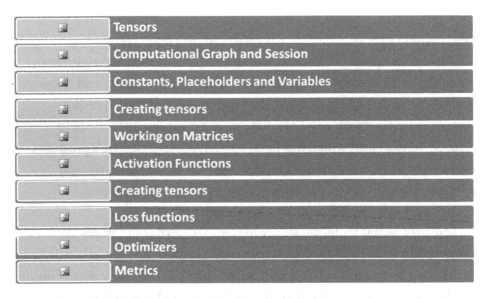

	Tensors
	Computational Graph and Session
	Constants, Placeholders and Variables
	Creating tensors
	Working on Matrices
	Activation Functions
	Creating tensors
	Loss functions
	Optimizers
	Metrics

Tensors

Before you jump into the TensorFlow library, let's get comfortable with the basic unit of data in TensorFlow. A *tensor* is a mathematical object and a generalization of scalars, vectors, and matrices. A tensor can be represented as a multidimensional array. A tensor of zero rank (order) is nothing but a scalar. A vector/array is a tensor of rank 1, whereas a matrix is a tensor of rank 2. In short, a tensor can be considered to be an *n*-dimensional array.

Here are some examples of tensors:

- 5: This is a rank 0 tensor; this is a scalar with shape [ ].

- [2.,5., 3.]: This is a rank 1 tensor; this is a vector with shape [3].

- [[1., 2., 7.], [3., 5., 4.]]: This is a rank 2 tensor; it is a matrix with shape [2, 3].

- [[[1., 2., 3.]], [[7., 8., 9.]]]: This is a rank 3 tensor with shape [2, 1, 3].

Computational Graph and Session

TensorFlow is popular for its TensorFlow Core programs where it has two main actions.

- Building the computational graph in the construction phase

- Running the computational graph in the execution phase

Let's understand how TensorFlow works.

- Its programs are usually structured into a construction phase and an execution phase.

- The construction phase assembles a graph that has nodes (ops/operations) and edges (tensors).

- The execution phase uses a session to execute ops (operations) in the graph.

3

- The simplest operation is a constant that takes no inputs but passes outputs to other operations that do computation.

- An example of an operation is multiplication (or addition or subtraction that takes two matrices as input and passes a matrix as output).

- The TensorFlow library has a default graph to which ops constructors add nodes.

So, the structure of TensorFlow programs has two phases, shown here:

Construction Phase
- Assembles a graph
- Constructs nodes(ops) and edges(tensors) of the computational graph

Execution Phase
- Uses a session to execute operations
- Repeatedly execute a set of training operations

A *computational graph* is a series of TensorFlow operations arranged into a graph of nodes.

Let's look at TensorFlow versus Numpy. In Numpy, if you plan to multiply two matrices, you create the matrices and multiply them. But in TensorFlow, you set up a graph (a default graph unless you create another graph). Next, you need to create variables, placeholders, and constant values and then create the session and initialize variables. Finally, you feed that data to placeholders so as to invoke any action.

To actually evaluate the nodes, you must run the computational graph within a session.

A *session* encapsulates the control and state of the TensorFlow runtime. The following code creates a Session object:

$$sess = tf.Session()$$

It then invokes its run method to run enough of the computational graph to evaluate node1 and node2.

The computation graph defines the computation. It neither computes anything nor holds any value. It is meant to define the operations mentioned in the code. A default graph is created. So, you don't need to create it unless you want to create graphs for multiple purposes.

A *session* allows you to execute graphs or parts of graphs. It allocates resources (on one or more CPUs or GPUs) for the execution. It holds the actual values of intermediate results and variables.

The value of a variable, created in TensorFlow, is valid only within one session. If you try to query the value afterward in a second session, TensorFlow will raise an error because the variable is not initialized there.

To run any operation, you need to create a session for that graph. The session will also allocate memory to store the current value of the variable

Here is the code to demonstrate:

```
import tensorflow as tf
sess = tf.Session()
```

```
# Creating a new graph(not default)
myGraph = tf.Graph()
with myGraph.as_default():
    variable = tf.Variable(30, name="navin")
    initialize = tf.global_variables_initializer()
```

```
with tf.Session(graph=myGraph) as sess:
    sess.run(initialize)
    print(sess.run(variable))
30
```

```
# Tensorboard can be used. It is optionalmy_
# Output graph can be seen on tensorboard
import os
merged = tf.summary.merge_all(key='summaries')
if not os.path.exists('tenosrboard_logs/'):
    os.makedirs('tenosrboard_logs/')

my_writer = tf.summary.FileWriter('/home/manaswi/tenosrboard_logs/', sess.graph)

def TB(cleanup=False):
    import webbrowser
    webbrowser.open('http://127.0.1.1:6006')
    !tensorboard --logdir='/home/manaswi/tenosrboard_logs'

    if cleanup:
        !rm -R tensorboard_logs/

TB(1)        # Launch graph on tensorborad on your browser
```

Constants, Placeholders, and Variables

TensorFlow programs use a tensor data structure to represent all data—
only tensors are passed between operations in the computation graph. You
can think of a TensorFlow tensor as an *n*-dimensional array or list. A tensor
has a static type, a rank, and a shape. Here the graph produces a constant
result. Variables maintain state across executions of the graph.

Generally, you have to deal with many images in deep learning, so you have to place pixel values for each image and keep iterating over all images.

To train the model, you need to be able to modify the graph to tune some objects such as weight and bias. In short, *variables* enable you to add trainable parameters to a graph. They are constructed with a type and initial value.

Let's create a constant in TensorFlow and print it.

```
import tensorflow as tf
x = tf.constant(12, dtype='float32')
sess = tf.Session()
print(sess.run(x))
```

12.0

Here is the explanation of the previous code in simple terms:

1. Import the `tensorflow` module and call it `tf`.

2. Create a constant value (x) and assign it the numerical value 12.

3. Create a session for computing the values.

4. Run just the variable x and print out its current value.

The first two steps belong to the construction phase, and the last two steps belong to the execution phase. I will discuss the construction and execution phases of TensorFlow now.

You can rewrite the previous code in another way, as shown here:

```
import tensorflow as tf
x = tf.constant(12, dtype='float32')
with tf.Session() as sess:
    print(sess.run(x))
```

12.0

Now you will explore how you create a variable and initialize it. Here is the code that does it:

```
import tensorflow as tf
x = tf.constant(12, dtype='float32')
y = tf.Variable(x+11)
model = tf.global_variables_initializer()
with tf.Session() as sess:
    sess.run(model)
    print(sess.run(y))
```

23.0

Here is the explanation of the previous code:

1. Import the `tensorflow` module and call it `tf`.

2. Create a constant value called x and give it the numerical value 12.

3. Create a variable called y and define it as being the equation 12+11.

4. Initialize the variables with `tf.global_variables_initializer()`.

5. Create a session for computing the values.

6. Run the model created in step 4.

7. Run just the variable y and print out its current value.

Here is some more code for your perusal:

```
import tensorflow as tf
x = tf.constant([14, 23, 40, 30])
y = tf.Variable(x*2 + 100)
model = tf.global_variables_initializer()
with tf.Session() as sess:
    sess.run(model)
    print(sess.run(y))
```

[128 146 180 160]

Placeholders

A *placeholder* is a variable that you can feed something to at a later time. It is meant to accept external inputs. Placeholders can have one or multiple dimensions, meant for storing *n*-dimensional arrays.

```
import tensorflow as tf
x = tf.placeholder("float", None)
y = x*10 + 500
with tf.Session() as sess:
    placeX = sess.run(y, feed_dict={x: [0, 5, 15, 25]})
    print(placeX)
```

```
[500. 550. 650. 750.]
```

Here is the explanation of the previous code:

1. Import the `tensorflow` module and call it `tf`.

2. Create a placeholder called x, mentioning the float type.

3. Create a tensor called y that is the operation of multiplying x by 10 and adding 500 to it. Note that any initial values for x are not defined.

4. Create a session for computing the values.

5. Define the values of x in `feed_dict` so as to run y.

6. Print out its value.

In the following example, you create a 2×4 matrix (a 2D array) for storing some numbers in it. You then use the same operation as before to do element-wise multiplying by 10 and adding 1 to it. The first dimension of the placeholder is None, which means any number of rows is allowed.

9

You can also consider a 2D array in place of the 1D array. Here is the code:

```
import tensorflow as tf
x = tf.placeholder("float", [None, 4])
y = x*10 + 1
with tf.Session() as sess:
    dataX = [[12, 2, 0, -2],
             [14, 4, 1, 0]]
    placeX = sess.run(y, feed_dict={x: dataX})
    print(placeX)
```
```
[[121.  21.   1. -19.]
 [141.  41.  11.   1.]]
```

This is a 2×4 matrix. So, if you replace None with 2, you can see the same output.

```
import tensorflow as tf
x = tf.placeholder("float", [2, 4])
y = x*10 + 1
with tf.Session() as sess:
    dataX = [[12, 2, 0, -2],
             [14, 4, 1, 0]]
    placeX = sess.run(y, feed_dict={x: dataX})
    print(placeX)
```
```
[[121.  21.   1. -19.]
 [141.  41.  11.   1.]]
```

But if you create a placeholder of [3, 4] shape (note that you will feed a 2×4 matrix at a later time), there is an error, as shown here:

```
import tensorflow as tf
x = tf.placeholder("float", [3, 4])
y = x*10 + 1
with tf.Session() as sess:
    dataX = [[12, 2, 0, -2],
             [14, 4, 1, 0]]
    placeX = sess.run(y, feed_dict={x: dataX})
    print(placeX)
```

```
------------------------------------------------------------------------
ValueError                                Traceback (most recent call last)
<ipython-input-10-c70a14b67e27> in <module>()
      5     dataX = [[12, 2, 0, -2],
      6              [14, 4, 1, 0]]
----> 7     placeX = sess.run(y, feed_dict={x: dataX})
      8     print(placeX)

~\Anaconda3\envs\tensorflow\lib\site-packages\tensorflow\python\client\session.py in run(self, fetches, feed_dict, options, run
_metadata)
    887         try:
    888             result = self._run(None, fetches, feed_dict, options_ptr,
--> 889                                run_metadata_ptr)
    890             if run_metadata:
    891                 proto_data = tf_session.TF_GetBuffer(run_metadata_ptr)

~\Anaconda3\envs\tensorflow\lib\site-packages\tensorflow\python\client\session.py in _run(self, handle, fetches, feed_dict, opt
ions, run_metadata)
   1094                     'Cannot feed value of shape %r for Tensor %r, '
   1095                     'which has shape %r'
-> 1096                     % (np_val.shape, subfeed_t.name, str(subfeed_t.get_shape())))
   1097                 if not self.graph.is_feedable(subfeed_t):
   1098                     raise ValueError('Tensor %s may not be fed.' % subfeed_t)

ValueError: Cannot feed value of shape (2, 4) for Tensor 'Placeholder_5:0', which has shape '(3, 4)'
```

```
################# What happens in a linear model ##########
# Weight and Bias as Variables as they are to be tuned
W = tf.Variable([2], dtype=tf.float32)
b = tf.Variable([3], dtype=tf.float32)
# Training dataset that will be fed while training as Placeholders
x = tf.placeholder(tf.float32)
# Linear Model
y = W * x + b
```

Constants are initialized when you call tf.constant, and their values can never change. By contrast, variables are not initialized when you call tf.Variable. To initialize all the variables in a TensorFlow program, you must explicitly call a special operation as follows.

```
sess.run(tf.global_variables_initializer())
```

It is important to realize init is a handle to the TensorFlow subgraph that initializes all the global variables. Until you call sess.run, the variables are uninitialized.

Creating Tensors

An image is a tensor of the third order where the dimensions belong to height, width, and number of channels (Red, Blue, and Green).

Here you can see how an image is converted into a tensor:

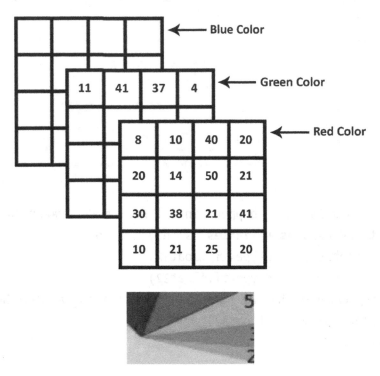

```
image = tf.image.decode_jpeg(tf.read_file("./Desktop/image.jpg"), channels=3)
sess = tf.InteractiveSession()
print(sess.run(tf.shape(image)))
```

```
[218 178    3]
```

```
print(sess.run(image[10:15,0:4,1]))
```

```
[[47 48 48 47]
 [45 45 45 44]
 [43 43 43 42]
 [41 42 42 41]
 [41 41 41 40]]
```

You can generate tensors of various types such as fixed tensors, random tensors, and sequential tensors.

Fixed Tensors

Here is a fixed tensor:

```
import tensorflow as tf
sess = tf.Session()
A = tf.zeros([2,3])
print(sess.run(A))
```

```
[[0. 0. 0.]
 [0. 0. 0.]]
```

```
B = tf.ones([4,3])
print(sess.run(B))
```

```
[[1. 1. 1.]
 [1. 1. 1.]
 [1. 1. 1.]
 [1. 1. 1.]]
```

```
import tensorflow as tf
sess = tf.Session()
A = tf.zeros([2,3])
print(sess.run(A))
```

```
[[0. 0. 0.]
 [0. 0. 0.]]
```
2 Rows
3 Columns

```
B = tf.ones([4,3])
print(sess.run(B))
```

```
[[1. 1. 1.]
 [1. 1. 1.]
 [1. 1. 1.]
 [1. 1. 1.]]
```
4 Rows
3 Columns

tf:.fill creates a tensor of shape (2×3) having a unique number.

```
C = tf.fill([2,3], 13)
print(sess.run(C))
```

```
[[13 13 13]
 [13 13 13]]
```

tf.diag creates a diagonal matrix having specified diagonal elements.

```
D = tf.diag([4,-3,2])
print(sess.run(D))
```

```
[[ 4  0  0]
 [ 0 -3  0]
 [ 0  0  2]]
```

tf.constant creates a constant tensor.

```
E = tf.constant([5,2,4,2])
print(sess.run(E))
```

```
[5 2 4 2]
```

Sequence Tensors

tf.range creates a sequence of numbers starting from the specified value and having a specified increment.

```
G = tf.range(start=6, limit=45, delta=3)
print(sess.run(G))
```

```
[ 6  9 12 15 18 21 24 27 30 33 36 39 42]
```

tf.linspace creates a sequence of evenly spaced values.

```
H = tf.linspace(10.0, 92.0, 5)
print(sess.run(H))
```

```
[10.   30.5 51.   71.5 92. ]
```

Random Tensors

`tf.random_uniform` generates random values from uniform distribution within a range.

```
R1 = tf.random_uniform([2,3], minval=0, maxval=4)
print(sess.run(R1))

[[0.74450636 1.9570832  3.1126966 ]
 [2.359518   2.101438   2.65689   ]]
```

`tf.random_normal` generates random values from the normal distribution having the specified mean and standard deviation.

```
R2 = tf.random_normal([2,3], mean=5, stddev=4)
print(sess.run(R2))

[[-1.8996243 3.2514744 5.9602127]
 [ 8.307009  4.84437   6.8460846]]
```

```
print(sess.run(tf.diag([3,-2,4])))

[[ 3  0  0]
 [ 0 -2  0]
 [ 0  0  4]]
```

```
R3 = tf.random_shuffle(tf.diag([3,-2,4]))
print(sess.run(R3))

[[ 3  0  0]
 [ 0  0  4]
 [ 0 -2  0]]
```

```
R4 = tf.random_crop(tf.diag([3,-2,4]), [3,2])
print(sess.run(R4))

[[ 3  0]
 [ 0 -2]
 [ 0  0]]
```

Can you guess the result?

```
print(sess.run(tf.zeros([2,4])))
```

```
print(sess.run(tf.diag([3,1,5,-2])))
```

```
print(sess.run(tf.range(start=4, limit=16, delta=2)))
```

If you are not able to find the result, please revise the previous portion where I discuss the creation of tensors.

Here you can see the result:

```
print(sess.run(tf.zeros([2,4])))
[[0. 0. 0. 0.]
 [0. 0. 0. 0.]]
```

```
print(sess.run(tf.diag([3,1,5,-2])))
[[ 3  0  0  0]
 [ 0  1  0  0]
 [ 0  0  5  0]
 [ 0  0  0 -2]]
```

```
print(sess.run(tf.range(start=4, limit=16, delta=2)))
[ 4  6  8 10 12 14]
```

Working on Matrices

Once you are comfortable creating tensors, you can enjoy working on matrices (2D tensors).

```
import tensorflow as tf
import numpy as np
sess = tf.Session()
A = tf.random_uniform([3,2])
B = tf.fill([2,4], 3.5)
C = tf.random_normal([3,4])
```

```
print(sess.run(A))
```

```
[[0.31633115 0.71407604]
 [0.18088198 0.36230946]
 [0.34481096 0.6156665 ]]
```

```
print(sess.run(B))
```

```
[[3.5 3.5 3.5 3.5]
 [3.5 3.5 3.5 3.5]]
```

```
print(sess.run(tf.matmul(A,B)))# Multiplication of Matrices
```

```
[[0.9453191 0.9453191 0.9453191 0.9453191]
 [4.4488316 4.4488316 4.4488316 4.4488316]
 [3.308284  3.308284  3.308284  3.308284 ]]
```

```
print(sess.run(tf.matmul(A,B) + C))# Multiplication & addition
```

```
[[4.6027136 4.5958595 6.9527874 4.413632 ]
 [3.3000264 4.4702578 5.0858393 5.168917 ]
 [3.1176403 4.626109  4.1446424 3.7285264]]
```

Activation Functions

The idea of an activation function comes from the analysis of how a neuron works in the human brain (see Figure 1-1). The neuron becomes active beyond a certain threshold, better known as the *activation potential*. It also attempts to put the output into a small range in most cases.

Sigmoid, hyperbolic tangent (tanh), ReLU, and ELU are most popular activation functions.

Let's look at the popular activation functions.

Tangent Hyperbolic and Sigmoid

Figure 1-2 shows the tangent hyperbolic and sigmoid activation functions.

Demonstration of Activation Function

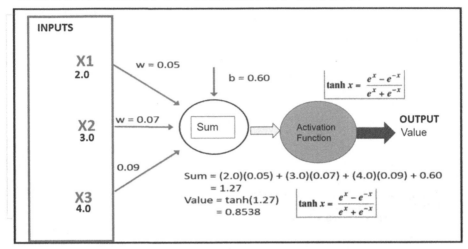

Figure 1-1. *An activation function*

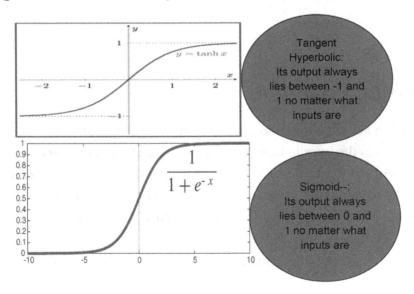

Figure 1-2. *Two popular activation functions*

Here is the demo code:

```
E = tf.nn.tanh([10,2,1,0.5,0,-0.5,-1.,-2.,-10.])
print(sess.run(E))
```

```
[ 1.          0.9640276   0.7615942    0.46211717  0.          -0.46211717
 -0.7615942  -0.9640276  -1.          ]
```

```
J = tf.nn.sigmoid([10,2,1,0.5,0,-0.5,-1.,-2.,-10.])
print(sess.run(J))
```

```
[9.9995458e-01 8.8079703e-01 7.3105860e-01 6.2245935e-01 5.0000000e-01
 3.7754068e-01 2.6894143e-01 1.1920292e-01 4.5397872e-05]
```

ReLU and ELU

Figure 1-3 shows the ReLU and ELU functions.

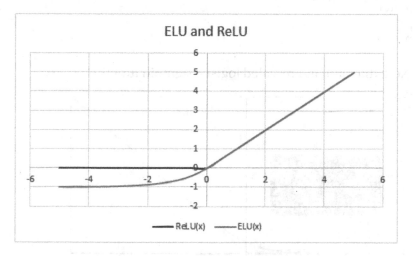

Figure 1-3. *The ReLU and ELU functions*

Here is the code to produce these functions:

```
A = tf.nn.relu([-2,1,-3,13])
print(sess.run(A))
```

```
[ 0  1  0 13]
```

ReLU6

ReLU6 is similar to ReLU except that the output cannot be more than six ever.

```
B = tf.nn.relu6([-2,1,-3,13])
print(sess.run(B))
```

```
[0 1 0 6]
```

```
C = tf.nn.relu([[-2,1,-3],[10,-16,-5]])
print(sess.run(C))
```

```
[[ 0  1  0]
 [10  0  0]]
```

Note that tanh is a rescaled logistic sigmoid function.

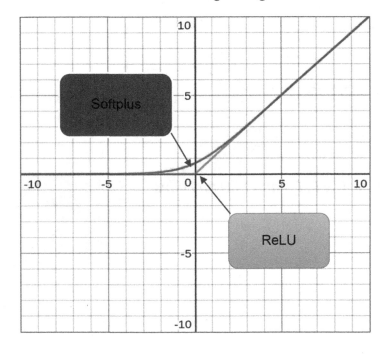

```
K = tf.nn.relu([10,2,1,0.5,0,-0.5,-1.,-2.,-10.])
print(sess.run(K))
```

```
[10.   2.   1.   0.5 0.   0.   0.   0.   0. ]
```

```
M = tf.nn.softplus([10,2,1,0.5,0,-0.5,-1.,-2.,-10.])
print(sess.run(M))
```

```
[1.0000046e+01 2.1269281e+00 1.3132616e+00 9.7407699e-01 6.9314718e-01
 4.7407699e-01 3.1326163e-01 1.2692805e-01 4.5417706e-05]
```

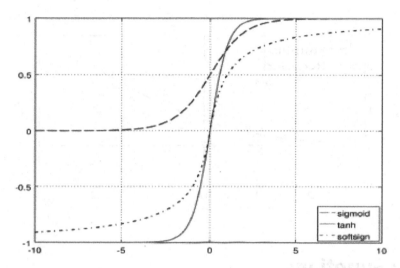

```
H = tf.nn.elu([10,2,1,0.5,0,-0.5,-1.,-2.,-10.])
print(sess.run(H))
```

```
[10.          2.          1.          0.5         0.          -0.39346933
 -0.63212055 -0.86466473 -0.9999546 ]
```

```
I = tf.nn.relu6([10,2,1,0.5,0,-0.5,-1.,-2.,-10.])
print(sess.run(I))
```

```
[6.  2.  1.  0.5 0.  0.  0.  0.  0. ]
```

21

```
G = tf.nn.softsign([10,2,1,0.5,0,-0.5,-1.,-2.,-10.])
print(sess.run(G))
```

```
[ 0.90909094  0.6666667   0.5         0.33333334  0.          -0.33333334
 -0.5        -0.6666667  -0.90909094]
```

```
F = tf.nn.softplus([10,2,1,0.5,0,-0.5,-1.,-2.,-10.])
print(sess.run(F))
```

```
[1.0000046e+01 2.1269281e+00 1.3132616e+00 9.7407699e-01 6.9314718e-01
 4.7407699e-01 3.1326163e-01 1.2692805e-01 4.5417706e-05]
```

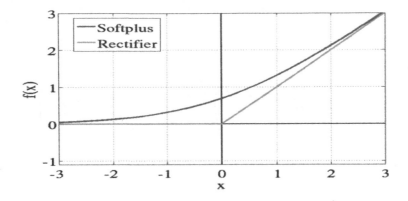

Loss Functions

The loss function (cost function) is to be minimized so as to get the best values for each parameter of the model. For example, you need to get the best value of the weight (slope) and bias (y-intercept) so as to explain the target (y) in terms of the predictor (X). The method is to achieve the best value of the slope, and y-intercept is to minimize the cost function/loss function/sum of squares. For any model, there are numerous parameters, and the model structure in prediction or classification is expressed in terms of the values of the parameters.

You need to evaluate your model, and for that you need to define the cost function (loss function). The minimization of the loss function can be the driving force for finding the optimum value of each parameter. For regression/numeric prediction, L1 or L2 can be the useful loss function. For classification, cross entropy can be the useful loss function. Softmax or sigmoid cross entropy can be quite popular loss functions.

Loss Function Examples

Here is the code to demonstrate:

```
import tensorflow as tf
import numpy as np
sess = tf.Session()
```

```
#Assuming prediction model
pred=np.asarray([0.2,0.3,0.5,10.0,12.0,13.0,3.5,7.4,3.9,2.3])
#convert ndarray into tensor
x_val=tf.convert_to_tensor(pred)
#Assuming actual values
actual=np.asarray([0.1,0.4,0.6,9.0,11.0,12.0,3.4,7.1,3.8,2.0])
```

```
#L2 loss:L1=(pred-actual)^2
l2=tf.square(pred-actual)
l2_out=sess.run(tf.round(l2))
print(l2_out)
```

```
[ 0.  0.  0.  1.  1.  1.  0.  0.  0.  0.]
```

```
#L2 loss:L1=abs(pred-actual)
l1=tf.abs(pred-actual)
l1_out=sess.run(l1)
print(l1_out)
```

```
[ 0.1  0.1  0.1  1.  1.  1.  0.1  0.3  0.1  0.3]
```

```
#cross entropy loss
softmax_xentropy_variable=tf.nn.sigmoid_cross_entropy_with_logits(logits=l1_out,labels=l2_out)
print(sess.run(softmax_xentropy_variable))
```

```
[ 0.74439666  0.74439666  0.74439666  0.31326169  0.31326169  0.31326169
  0.74439666  0.85435524  0.74439666  0.85435524]
```

Common Loss Functions

The following is a list of the most common loss functions:

tf.contrib.losses.absolute_difference

tf.contrib.losses.add_loss

```
tf.contrib.losses.hinge_loss

tf.contrib.losses.compute_weighted_loss

tf.contrib.losses.cosine_distance

tf.contrib.losses.get_losses

tf.contrib.losses.get_regularization_losses

tf.contrib.losses.get_total_loss

tf.contrib.losses.log_loss

tf.contrib.losses.mean_pairwise_squared_error

tf.contrib.losses.mean_squared_error

tf.contrib.losses.sigmoid_cross_entropy

tf.contrib.losses.softmax_cross_entropy

tf.contrib.losses.sparse_softmax_cross_entropy

tf.contrib.losses.log(predictions,labels,weight=2.0)
```

Optimizers

Now you should be convinced that you need to use a loss function to get the best value of each parameter of the model. How can you get the best value?

Initially you assume the initial values of weight and bias for the model (linear regression, etc.). Now you need to find the way to reach to the best value of the parameters. The optimizer is the way to reach the best value of the parameters. In each iteration, the value changes in a direction suggested by the optimizer. Suppose you have 16 weight values (w1, w2, w3, ..., w16) and 4 biases (b1, b2, b3, b4). Initially you can assume every weight and bias to be zero (or one or any number). The optimizer suggests whether w1 (and other parameters) should increase or decrease in the next iteration while keeping the goal of minimization in mind. After many iterations, w1 (and other parameters) would stabilize to the best value (or values) of parameters.

In other words, TensorFlow, and every other deep learning framework, provides optimizers that slowly change each parameter in order to minimize the loss function. The purpose of the optimizers is to give direction to the weight and bias for the change in the next iteration. Assume that you have 64 weights and 16 biases; you try to change the weight and bias values in each iteration (during backpropagation) so that you get the correct values of weights and biases after many iterations while trying to minimize the loss function.

Selecting the best optimizer for the model to converge fast and to learn weights and biases properly is a tricky task.

Adaptive techniques (adadelta, adagrad, etc.) are good optimizers for converging faster for complex neural networks. Adam is supposedly the best optimizer for most cases. It also outperforms other adaptive techniques (adadelta, adagrad, etc.), but it is computationally costly. For sparse data sets, methods such as SGD, NAG, and momentum are not the best options; the *adaptive learning rate* methods are. An additional benefit

25

is that you won't need to adjust the learning rate but can likely achieve the best results with the default value.

Loss Function Examples

Here is the code to demonstrate:

```
# Importing libraries
import tensorflow as tf
```

```
# Assign the value into variable
x = tf.Variable(3, name='x', dtype=tf.float32)
log_x = tf.log(x)
log_x_squared = tf.square(log_x)
```

```
# Apply GradientDescentOptimizer
optimizer = tf.train.GradientDescentOptimizer(0.7)
train = optimizer.minimize(log_x_squared)
```

```
# Initialize Variables
init = tf.global_variables_initializer()
```

```
# Finally running computation
with tf.Session() as session:
    session.run(init)
    print("starting at", "x:", session.run(x), "log(x)^2:", session.run(log_x_squared))
    for step in range(10):
        session.run(train)
        print("step", step, "x:", session.run(x), "log(x)^2:", session.run(log_x_squared))
```

```
starting at x: 3.0 log(x)^2: 1.20695
step 0 x: 2.48731 log(x)^2: 0.830292
step 1 x: 1.97444 log(x)^2: 0.462786
step 2 x: 1.49207 log(x)^2: 0.160134
step 3 x: 1.1166 log(x)^2: 0.0121637
step 4 x: 0.97832 log(x)^2: 0.00048043
step 5 x: 1.00969 log(x)^2: 9.29177e-05
step 6 x: 0.99632 log(x)^2: 1.35901e-05
step 7 x: 1.0015 log(x)^2: 2.24809e-06
step 8 x: 0.999405 log(x)^2: 3.54772e-07
step 9 x: 1.00024 log(x)^2: 5.70574e-08
```

Common Optimizers

The following is a list of common optimizers:

tf.train.Optimizer
tf.train.GradientDescentOptimizer
tf.train.AdadeltaOptimizer
tf.train.AdagradOptimizer
tf.train.AdagradDAOptimizer
tf.train.MomentumOptimizer
tf.train.AdamOptimizer
tf.train.FtrlOptimizer
tf.train.ProximalGradientDescentOptimizer
tf.train.ProximalAdagradOptimizer
tf.train.RMSPropOptimizer

Metrics

Having learned some ways to build a model, it is time to evaluate the model. So, you need to evaluate the regressor or classifier.

There are many evaluation metrics, among which classification accuracy, logarithmic loss, and area under ROC curve are the most popular ones.

Classification accuracy is the ratio of the number of correct predictions to the number of all predictions. When observations for each class are not much skewed, accuracy can be considered as a good metric.

```
tf.contrib.metrics.accuracy(actual_labels, predictions)
```

There are other evaluation metrics as well.

Metrics Examples

This section shows the code to demonstrate.

Here you create actual values (calling them x) and predicted values (calling them y). Then you check the accuracy. Accuracy represents the ratio of the number of times the actual equals the predicted values and total number of instances.

Common Metrics

The following is a list of common metrics:

tf.contrib.metrics.streaming_root_mean_squared_error
tf.contrib.metrics.streaming_covariance
tf.contrib.metrics.streaming_pearson_correlation
tf.contrib.metrics.streaming_mean_cosine_distance
tf.contrib.metrics.streaming_percentage_less
tf.contrib.metrics.streaming_sensitivity_at_specificity
tf.contrib.metrics.streaming_sparse_average_precision_at_k
tf.contrib.metrics.streaming_sparse_precision_at_k tf.contrib.metrics.streaming_sparse_precision_at_top_k
tf.contrib.metrics.streaming_specificity_at_sensitivity
tf.contrib.metrics.streaming_concat
tf.contrib.metrics.streaming_false_negatives
tf.contrib.metrics.streaming_false_negatives_at_thresholds

```
# Importing libraries
import numpy as np
import tensorflow as tf
```

```
# Placeholders declaration
x=tf. placeholder(tf.int32, [5])
y=tf. placeholder(tf.int32, [5])
```

```
# Metrices declaration
acc, acc_op=tf.metrics.accuracy(labels=x, predictions=y)
```

```
# Session initialization
sess=tf.InteractiveSession()
sess.run(tf.global_variables_initializer())
sess.run(tf.local_variables_initializer())
```

```
# Value assign
val= sess.run([acc,acc_op], feed_dict={x: [1,1,0,1,0], y: [0,1,0,0,1]})
```

```
# Print Accuracy
val_acc=sess.run(acc)
print(val_acc)
# You can see only 2nd and 3rd positions value are same
```

0.4

CHAPTER 2

Understanding and Working with Keras

Keras is a compact and easy-to-learn high-level Python library for deep learning that can run on top of TensorFlow (or Theano or CNTK). It allows developers to focus on the main concepts of deep learning, such as creating layers for neural networks, while taking care of the nitty-gritty details of tensors, their shapes, and their mathematical details. TensorFlow (or Theano or CNTK) has to be the back end for Keras. You can use Keras for deep learning applications without interacting with the relatively complex TensorFlow (or Theano or CNTK). There are two major kinds of framework: the sequential API and the functional API. The sequential API is based on the idea of a sequence of layers; this is the most common usage of Keras and the easiest part of Keras. The sequential model can be considered as a linear stack of layers.

In short, you create a sequential model where you can easily add layers, and each layer can have convolution, max pooling, activation, drop-out, and batch normalization. Let's go through major steps to develop deep learning models in Keras.

© Navin Kumar Manaswi 2018
N. K. Manaswi, *Deep Learning with Applications Using Python*,
https://doi.org/10.1007/978-1-4842-3516-4_2

Major Steps to Deep Learning Models

The four core parts of deep learning models in Keras are as follows:

1. Define the model. Here you create a sequential model and add layers. Each layer can contain one or more convolution, pooling, batch normalization, and activation function.

2. Compile the model. Here you apply the loss function and optimizer before calling the `compile()` function on the model.

3. Fit the model with training data. Here you train the model on the test data by calling the `fit()` function on the model.

4. Make predictions. Here you use the model to generate predictions on new data by calling functions such as `evaluate()` and `predict()`.

There are eight steps to the deep learning process in Keras:

1. Load the data.

2. Preprocess the data.

3. Define the model.

4. Compile the model.

5. Fit the model.

6. Evaluate the model.

7. Make the predictions.

8. Save the model.

Load Data

Here is how you load data:

```
# Importing modules
import numpy as np
import os
from keras.datasets import cifar10
from keras.models import Sequential
from keras.layers.core import Dense, Dropout, Activation
from keras.optimizers import adam
from keras.utils import np_utils
```

```
#Load Data
np.random.seed(100) # for reproducibility
(X_train, y_train), (X_test, y_test) = cifar10.load_data()

#cifar-10 has images of airplane, automobile, bird, cat,
# deer, dog, frog, horse, ship and truck ( 10 unique labels)
# For each image. width = 32, height =32, Number of channels(RGB) = 3
```

Preprocess the Data

Here is how you preprocess data:

```
#Preprocess the data
#Flatten the data, MLP doesn't use the 2D structure of the data. 3072 = 3*32*32
X_train = X_train.reshape(50000, 3072) # 50,000 images for training
X_test = X_test.reshape(10000, 3072) # 10,000 images for test

# Gaussian Normalization( Z- score)
X_train = (X_train- np.mean(X_train))/np.std(X_train)
X_test = (X_test- np.mean(X_test))/np.std(X_test)
```

```
# Convert class vectors to binary class matrices (ie one-hot vectors)
labels = 10 #10 unique labels(0-9)
Y_train = np_utils.to_categorical(y_train, labels)
Y_test = np_utils.to_categorical(y_test, labels)
```

Define the Model

Sequential models in Keras are defined as a sequence of layers. You create a sequential model and then add layers. You need to ensure the input layer has the right number of inputs. Assume that you have 3,072 input variables; then you need to create the first hidden layer with 512 nodes/neurons. In the second hidden layer, you have 120 nodes/neurons. Finally, you have ten nodes in the output layer. For example, an image maps onto ten nodes that shows the probability of being label1 (airplane), label2 (automobile), label3 (cat), ..., label10 (truck). The node of highest probability is the predicted class/label.

```
#Define the model achitecture
model = Sequential()
model.add(Dense(512, input_shape=(3072,))) # 3*32*32 = 3072
model.add(Activation('relu'))
model.add(Dropout(0.4)) # Regularization
model.add(Dense(120))
model.add(Activation('relu'))
model.add(Dropout(0.2))# Regularization
model.add(Dense(labels)) #Last Layer with 10 outputs, each output per class
model.add(Activation('sigmoid'))
```

One image has three channels (RGB), and in each channel, the image has 32×32 = 1024 pixels. So, each image has 3×1024 = 3072 pixels (features/X/inputs).

With the help of 3,072 features, you need to predict the probability of label1 (Digit 0), label2 (Digit 1), and so on. This means the model predicts ten outputs (Digits 0–9) where each output represents the probability of the corresponding label. The last activation function (sigmoid, as shown earlier) gives 0 for nine outputs and 1 for only one output. That label is the predicted class for the image (Figure 2-1).

For example, 3,072 features ➤ 512 nodes ➤ 120 nodes ➤ 10 nodes.

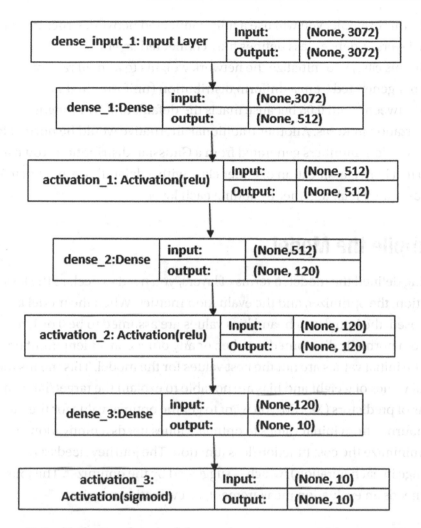

Figure 2-1. *Defining the model*

The next question is, how do you know the number of layers to use and their types? No one has the exact answer. What's best for evaluation metrics is that you decide the optimum number of layers and the parameters and steps in each layer. A heuristics approach is also used. The best network structure is found through a process of trial-and-error experimentation. Generally, you need a network large enough to capture the structure of the problem.

In this example, you will use a fully connected network structure with three layers. A dense class defines fully connected layers.

In this case, you initialize the network weights to a small random number generated from a uniform distribution (uniform) in this case between 0 and 0.05 because that is the default uniform weight initialization in Keras. Another traditional alternative would be normal for small random numbers generated from a Gaussian distribution. You use or snap to a hard classification of either class with a default threshold of 0.5. You can piece it all together by adding each layer.

Compile the Model

Having defined the model in terms of layers, you need to declare the loss function, the optimizer, and the evaluation metrics. When the model is proposed, the initial weight and bias values are assumed to be 0 or 1, a random normally distributed number, or any other convenient numbers. But the initial values are not the best values for the model. This means the initial values of weight and bias are not able to explain the target/label in terms of predictors (Xs). So, you want to get the optimal value for the model. The journey from initial values to optimal values needs a motivation, which will minimize the cost function/loss function. The journey needs a path (change in each iteration), which is suggested by the optimizer. The journey also needs an evaluation measurement, or evaluation metrics.

```
# Compile the model
# Use adam as an optimizer
adam = adam(0.01)
# the cross entropy between the true label and the output(softmax) of the model
model.compile(loss='categorical_crossentropy', optimizer=adam, metrics=["accuracy"])
```

Popular loss functions are binary cross entropy, categorical cross entropy, mean_squared_logarithmic_error and hinge loss. Popular optimizers are stochastic gradient descent (SGD), RMSProp, adam, adagrad, and adadelta. Popular evaluation metrics are accuracy, recall, and F1 score.

In short, this step is aimed at tuning the weights and biases based on loss functions through iterations based on the optimizer evaluated by metrics such as accuracy.

Fit the Model

Having defined and compiled the model, you need to make predications by executing the model on some data. Here you need to specify the epochs; these are the number of iterations for the training process to run through the data set and the batch size, which is the number of instances that are evaluated before a weight update. For this problem, the program will run for a small number of epochs (10), and in each epoch, it will complete 50(=50,000/1,000) iterations where the batch size is 1,000 and the training data set has 50,000 instances/images. Again, there is no hard rule to select the batch size. But it should not be very small, and it should be much less than the size of the training data set to consume less memory.

```
#Make the model learn ( Fit the model)
model.fit(X_train, Y_train,batch_size=1000, nb_epoch=10,validation_data=(X_test, Y_test))

Train on 50000 samples, validate on 10000 samples
Epoch 1/10
 1000/50000 [..............................] - ETA: 6s - loss: 2.3028 - acc: 0.1060
C:\ProgramData\Anaconda3\lib\site-packages\keras\models.py:848: UserWarning: The `nb_epoch` argument in `fit` has been renamed `epochs`.
  warnings.warn('The `nb_epoch` argument in `fit` '
50000/50000 [==============================] - 6s - loss: 2.3030 - acc: 0.0974 - val_loss: 2.3027 - val_acc: 0.1000
Epoch 2/10
50000/50000 [==============================] - 7s - loss: 2.3029 - acc: 0.1012 - val_loss: 2.3027 - val_acc: 0.1000
Epoch 3/10
50000/50000 [==============================] - 7s - loss: 2.3028 - acc: 0.0972 - val_loss: 2.3026 - val_acc: 0.1000
Epoch 4/10
50000/50000 [==============================] - 7s - loss: 2.3028 - acc: 0.0997 - val_loss: 2.3027 - val_acc: 0.1000
Epoch 5/10
50000/50000 [==============================] - 7s - loss: 2.3029 - acc: 0.0975 - val_loss: 2.3027 - val_acc: 0.1000
Epoch 6/10
50000/50000 [==============================] - 7s - loss: 2.3029 - acc: 0.0986 - val_loss: 2.3028 - val_acc: 0.1000
Epoch 7/10
50000/50000 [==============================] - 7s - loss: 2.3029 - acc: 0.0995 - val_loss: 2.3028 - val_acc: 0.1000
Epoch 8/10
50000/50000 [==============================] - 7s - loss: 2.3028 - acc: 0.0983 - val_loss: 2.3027 - val_acc: 0.1000
Epoch 9/10
50000/50000 [==============================] - 7s - loss: 2.3029 - acc: 0.0998 - val_loss: 2.3027 - val_acc: 0.1000
Epoch 10/10
50000/50000 [==============================] - 7s - loss: 2.3029 - acc: 0.0972 - val_loss: 2.3027 - val_acc: 0.1000

<keras.callbacks.History at 0x2870136eef0>
```

Evaluate Model

Having trained the neural networks on the training data sets, you need to evaluate the performance of the network. Note that this will only give you an idea of how well you have modeled the data set (e.g., the train accuracy), but you won't know how well the algorithm might perform on new data. This is for simplicity, but ideally, you could separate your data into train and test data sets for the training and evaluation of your model. You can evaluate your model on your training data set using the evaluation() function on your model and pass it the same input and output used to train the model. This will generate a prediction for each input and output pair and collect scores, including the average loss and any metrics you have configured, such as accuracy.

```
#Evaluate how the model does on the test set
score = model.evaluate(X_test, Y_test, verbose=0)
#Accuracy Score
print('Test accuracy:', score[1])
```

Prediction

Once you have built and evaluated the model, you need to predict for unknown data.

```
#Predict digit(0-9) for test Data
model.predict_classes(X_test)
```

```
 9888/10000 [=============================>.] - ETA: 0s

array([3, 8, 8, ..., 3, 4, 7], dtype=int64)
```

Save and Reload the Model

Here is the final step:

```
#Saving the model
model.save('model.h5')
jsonModel = model.to_json()
model.save_weights('modelWeight.h5')
```

```
#Load weight of the saved model
modelWt = model.load_weights('modelWeight.h5')
```

Optional: Summarize the Model

Now let's see how to summarize the model.

```
#Summary of the model
model.summary()
```

Layer (type)	Output Shape	Param #
dense_7 (Dense)	(None, 512)	1573376
activation_7 (Activation)	(None, 512)	0
dropout_5 (Dropout)	(None, 512)	0
dense_8 (Dense)	(None, 120)	61560
activation_8 (Activation)	(None, 120)	0
dropout_6 (Dropout)	(None, 120)	0
dense_9 (Dense)	(None, 10)	1210
activation_9 (Activation)	(None, 10)	0

```
Total params: 1,636,146
Trainable params: 1,636,146
Non-trainable params: 0
```

Additional Steps to Improve Keras Models

Here are some more steps to improve your models:

1. Sometimes, the model building process does not complete because of a vanishing or exploding gradient. If this is the case, you should do the following:

```
from keras.callbacks import EarlyStopping
early_stopping_ monitor = EarlyStopping(patience=2)
model.fit(x_train, y_train, batch_size=1000, epochs=10,
validation_data=(x_test, y_test),
callbacks=[early_stopping_monitor])
```

2. Model the output shape.

   #Shape of the n-dim array (output of the model at the current position)
   ```
   model.output_shape
   ```

3. Model the summary representation.

   ```
   model.summary()
   ```

4. Model the configuration.

   ```
   model.get_config()
   ```

5. List all the weight tensors in the model.

   ```
   model.get_weights()
   ```

Here I am sharing the complete code for the Keras model. Can you attempt to explain it?

A TYPING DEEP LEARNING MODEL WITH KERAS

```
import numpy as np
from keras.models import Sequential
from keras.layers import Dense
```

Loading Data

```
data = np.random.random((500,100))
labels = np.random.randint(2,size=(500,1))
```

Create model

```
model = Sequential()
model.add(Dense(12, input_dim=8, activation='relu'))
model.add(Dense(8, activation='relu'))
model.add(Dense(1, activation='sigmoid'))
```

Compile model

```
model.compile(loss='binary_crossentropy', optimizer='adam',
metrics=['accuracy'])
```

Fit the model

```
model.fit(X[train], Y[train], epochs=150, batch_size=10, verbose=0)
```

Evaluate the model

```
scores = model.evaluate(X[test], Y[test], verbose=0)
print("%s: %.2f%%" % (model.metrics_names[1], scores[1]*100))
cvscores.append(scores[1] * 100) print("%.2f%% (+/- %.2f%%)" %
(numpy.mean(cvscores), numpy.std(cvscores)))
```

Predict

```
predictions = model.predict(data)
```

Keras with TensorFlow

Keras provides high-level neural networks by leveraging a powerful and lucid deep learning library on top of TensorFlow/Theano. Keras is a great addition to TensorFlow as its layers and models are compatible with pure-TensorFlow tensors. Moreover, it can be used alongside other TensorFlow libraries.

Here are the steps involved in using Keras for TensorFlow:

1. Start by creating a TensorFlow session and registering it with Keras. This means Keras will use the session you registered to initialize all the variables that it creates internally.

```
import TensorFlow as tf
sess = tf.Session()
from keras import backend as K
K.set_session(sess)
```

2. Keras modules such as the model, layers, and activation are used to build models. The Keras engine automatically converts these modules into the TensorFlow-equivalent script.

3. Other than TensorFlow, Theano and CNTK can be used as back ends to Keras.

4. A TensorFlow back end has the convention of making the input shape (to the first layer of your network) in depth, height, width order, where depth can mean the number of channels.

5. You need to configure the keras.json file correctly
    so that it uses the TensorFlow back end. It should
    look something like this:

```
{
    "backend": "theano",
    "epsilon": 1e-07,
    "image_data_format": "channels_first",
    "floatx": "float32"
}
```

In next chapters, you will learn how to leverage Keras for working on CNN, RNN, LSTM, and other deep learning activities.

CHAPTER 3

Multilayer Perceptron

Before you start learning about multilayered perceptron, you need to get a big-picture view of artificial neural networks. That's what I'll start with in this chapter.

Artificial Neural Network

An *artificial neural network* (ANN) is a computational network (a system of nodes and the interconnection between nodes) inspired by biological neural networks, which are the complex networks of neurons in human brains (see Figure 3-1). The nodes created in the ANN are supposedly programmed to behave like actual neurons, and hence they are artificial neurons. Figure 3-1 shows the network of the nodes (artificial neurons) that make up the artificial neural network.

© Navin Kumar Manaswi 2018
N. K. Manaswi, *Deep Learning with Applications Using Python*,
https://doi.org/10.1007/978-1-4842-3516-4_3

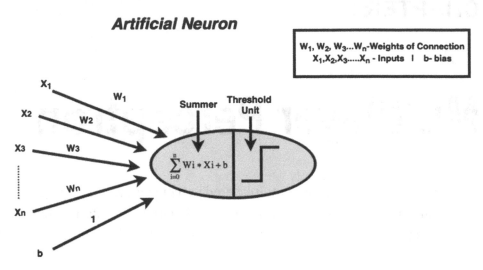

Figure 3-1. *Artificial neural network*

The number of layers and the number of neurons/nodes per layer can be the main structural component of an artificial neural network. Initially, the weights (representing the interconnection) and bias are not good enough to make the decision (classification, etc.). It is like the brain of a baby who has no prior experience. A baby learns from experiences so as to be a good decision-maker (classifier). Experiences/data (labeled) helps the neural network of brains tune the (neural) weights and bias. The artificial neural network goes through the same process. The weights are tuned per iteration to create a good classifier. Since tuning and thereby getting the correct weights by hand for thousands of neurons is very time-consuming, you use algorithms to perform these duties.

That process of tuning the weights is called *learning* or *training*. This is the same as what humans do on a daily basis. We try to enable computers to perform like humans.

Let's start exploring the simplest ANN model.

A typical neural network contains a large number of artificial neurons called *units* arranged in a series of different layers: input layer, hidden layer, and output layer (Figure 3-2).

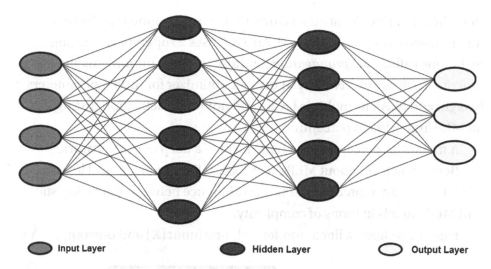

⬤ Input Layer	⬤ Hidden Layer	◯ Output Layer

Figure 3-2. *Neural network*

Neural networks are connected, which means each neuron in the hidden layer is fully connected to every neuron in the previous input layer and to its next output layer. A neural network learns by adjusting the weights and biases in each layer iteratively to get the optimal results.

Single-Layer Perceptron

A *single-layer perceptron* is a simple linear binary classifier. It takes inputs and associated weights and combines them to produce output that is used for classification. It has no hidden layers. Logistic regression is the single-layer perceptron.

Multilayer Perceptron

A *multilayer perceptron* (MLP) is a simple example of feedback artificial neural networks. An MLP consists of at least one hidden layer of nodes other than the input layer and the output layer. Each node of a layer other than the input layer is called a *neuron* that uses a nonlinear activation function such as sigmoid or ReLU. An MLP uses a supervised learning technique called *backpropagation* for training, while minimizing the loss function such as cross entropy. It uses an optimizer for tuning parameters (weight and bias). Its multiple layers and nonlinear activation distinguish an MLP from a linear perceptron.

A multilayer perceptron is a basic form of a deep neural network.

Before you learn about MLPs, let's look at linear models and logistic models. You can appreciate the subtle difference between linear, logistic, and MLP models in terms of complexity.

Figure 3-3 shows a linear model with one input (X) and one output (Y).

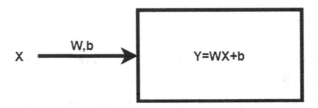

Figure 3-3. *Single-input vector*

The single-input model has a vector X with weight W and bias b. The output, Y, is WX + b, which is the linear model.

Figure 3-4 shows multiple inputs (X1 and X2) and one output (Y).

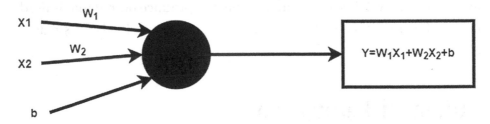

Figure 3-4. *Linear model*

This linear model has two input features: X1 and X2 with the corresponding weights to each input feature being W1, W2, and bias b. The output, Y, is W1X1 + W2X2 + b.

Logistic Regression Model

Figure 3-5 shows the learning algorithm that you use when the output label Y is either 0 or 1 for a binary classification problem. Given an input feature vector X, you want the probability that Y = 1 given the input feature X. This is also called as a *shallow* neural network or a *single-layer* (no hidden layer; only and output layer) neural network. The output layer, Y, is σ (Z), where Z is WX + b and σ is a sigmoid function.

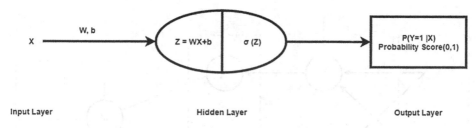

Figure 3-5. *One input (X) and one output (Y)*

Figure 3-6 shows the learning algorithm that you use when the output label Y is either 0 or 1 for a binary classification problem.

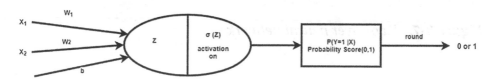

Figure 3-6. *Multiple inputs (X1 and X1) and one output (Y)*

49

Given input feature vectors X1 and X2, you want the probability that Y = 1 given the input features. This is also called a *perceptron*. The output layer, Y, is σ (Z), where Z is WX + b.

$$\begin{bmatrix} X1 \\ X2 \end{bmatrix} \rightarrow \begin{bmatrix} W1 & W2 \\ W3 & W4 \end{bmatrix}\begin{bmatrix} X1 \\ X2 \end{bmatrix} + \begin{bmatrix} b1 \\ b2 \end{bmatrix} \rightarrow \sigma\left(\begin{bmatrix} W1*X1+W2*X2+b1 \\ W3*X1+W4*X2+b2 \end{bmatrix} \right)$$

Figure 3-7 shows a two-layer neural network, with a hidden layer and an output layer. Consider that you have two input feature vectors X1 and X2 connecting to two neurons, X1' and X2.' The parameters (weights) associated from the input layer to the hidden layer are w1, w2, w3, w4, b1, b2.

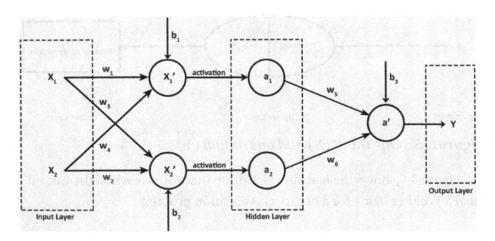

Figure 3-7. *Two-layer neural network*

X1' and X2' compute the linear combination (Figure 3-8).

$$\begin{bmatrix} X1' \\ X2' \end{bmatrix} = \begin{bmatrix} w1 & w2 \\ w3 & w4 \end{bmatrix} \begin{bmatrix} X1 \\ X2 \end{bmatrix} + \begin{bmatrix} b1 \\ b2 \end{bmatrix}$$

$(2 \times 1)(2 \times 2)(2 \times 1)(2 \times 1)$ is the dimension of the input and hidden layers.

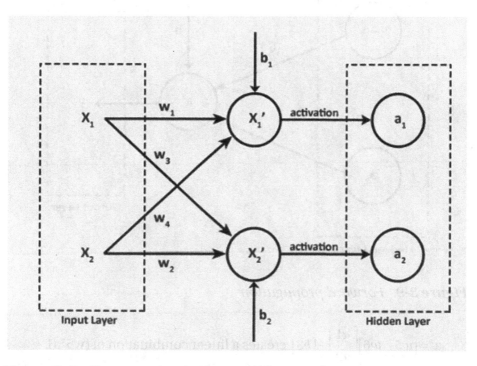

Figure 3-8. *Computation in the neural network*

The linear input X1' and X2' passes through the activation unit a1 and a2 in the hidden layer.

a1 is σ (X1') and a2 is σ(X2'), so you can also write the equation as follows:

$$\begin{bmatrix} a1 \\ a2 \end{bmatrix} = \sigma \begin{bmatrix} X1' \\ X2' \end{bmatrix}$$

The value forward propagates from the hidden layer to the output layer. Inputs a1 and a2 and parameters w5, w6, and b3 pass through the output layer a' (Figure 3-9).

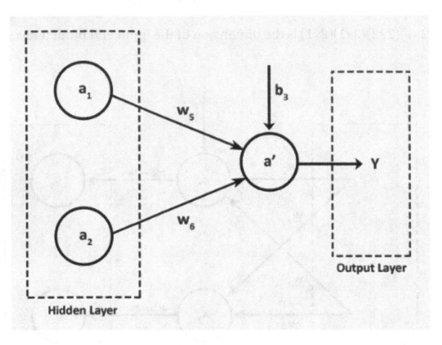

Figure 3-9. *Forward propagation*

$a' = \begin{bmatrix} w5 & w6 \end{bmatrix} \begin{bmatrix} a1 \\ a2 \end{bmatrix} + \begin{bmatrix} b3 \end{bmatrix}$ creates a linear combination of (w5*a1 + w6*a2) + b3, which will pass through a nonlinear sigmoid function to the final output layer, Y.

$$y = \sigma(a')$$

Let's say the initial model structure in one dimension is Y = w*X + b, where the parameters w and b are weights and bias.

Consider the loss function L(w, b) = 0.9 for the initial value of the parameters w = 1 and b = 1. You get this output: y = 1*X+1 & L(w ,b) = 0.9.

The objective is to minimize the loss by adjusting the parameters w and b. The errors will be backpropagated from the output layer to the hidden layer to the input layer to adjust the parameter through a learning rate and optimizer. Finally, we want to build a model (regressor) that can explain Y in terms of X.

To start the process of build a model, we initialize weight and bias. For convenience, w = 1, b = 1 (Initial value), (optimizer) stochastic gradient descent with learning rate (α = 0.01).

Here is step 1: Y = 1 * X + 1.

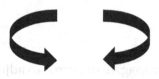

1.20   0.35

The parameters are adjusted to w = 1.20 and b = 0.35.

Here is step 2: Y1 = 1.20*X + 0.35.

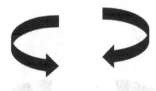

1.24   0.31

The parameters are adjusted to w = 1.24 and b = 0.31.

Here is step 3: Y1 = 1.24*X + 0.31.

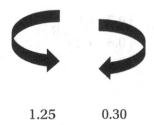

1.25   0.30

After some iterations, the weight and bias become stable. As you see, the initial changes are high while tuning. After some iterations, the change is not significant.

L(w, b) gets minimized for w = 1.26 and b = 0.29; hence, the final model becomes the following:

Y = 1.26 * X + 0.29

Similarly, in two dimensions, you can consider the parameters, weight matrix and bias vector.

Let's assume that initial weight matrix and bias vector as $W = \begin{bmatrix} 1 & 1 \\ 1 & 1 \end{bmatrix}$ and $B = \begin{bmatrix} 1 \\ 1 \end{bmatrix}$.

You iterate and backpropagate the error to adjust w and b.

$Y = W = \begin{bmatrix} 1 & 1 \\ 1 & 1 \end{bmatrix} * [X] + \begin{bmatrix} 1 \\ 1 \end{bmatrix}$ is the initial model. Weight matrix (2x2) and bias matrix(2x1) are tuned in each iteration. So, we can see change in weight and bias matrices

Here is step 1:

$$W = \begin{bmatrix} 0.7 & 0.8 \\ 0.6 & 1.2 \end{bmatrix}, B = \begin{bmatrix} 2.4 \\ 3.2 \end{bmatrix}$$

Here is step 2:

$$\begin{bmatrix} 0.7 & 0.8 \\ 0.6 & 1.2 \end{bmatrix} \begin{bmatrix} 2.4 \\ 3.2 \end{bmatrix}$$

$$W = \begin{bmatrix} 0.6 & 0.7 \\ 0.4 & 1.3 \end{bmatrix}, B = \begin{bmatrix} 2.8 \\ 3.8 \end{bmatrix}$$

Here is step 3:

$$\begin{bmatrix} 0.6 & 0.7 \\ 0.4 & 1.3 \end{bmatrix} \begin{bmatrix} 2.8 \\ 3.8 \end{bmatrix}$$

You can notice change in weight matrix(2x2) and bias matrix(2x1) in the iteration.

$$W = \begin{bmatrix} 0.5 & 0.6 \\ 0.3 & 1.3 \end{bmatrix}, B = \begin{bmatrix} 2.9 \\ 4.0 \end{bmatrix}$$

The final model after w and b are adjusted is as follows:

$$Y = \begin{bmatrix} 0.4 & 0.5 \\ 0.2 & 1.3 \end{bmatrix} * [X] + \begin{bmatrix} 3.0 \\ 4.0 \end{bmatrix}$$

In this chapter, you learned how weight and bias are tuned in each iteration while keeping the aim of minimization of loss functions. That is done with the help of optimizers such as stochastic gradient descent.

In this chapter, we have understood ANN and MLP as the basic deep learning model. Here, we can see MLP as the natural progression from linear and logistic regression. We have seen how weight and bias are tuned in every iteration which happens in backpropagation. Without going into details of backpropagation, we have seen the action/result of backpropagation. In next two chapters, we can learn how to build MLP models in TensorFlow and in keras.

CHAPTER 4

Regression to MLP in TensorFlow

People have been using regression and classifiers for a long time. Now it is time to switch to the topic of neural networks. A *multilayered perceptron* (MLP) is a simple neural network model where you can add one or more hidden layers between the input and output layers.

In this chapter, you will see how TensorFlow can help you build the models. You will start with the most basic model, which is a linear model. Logistic and MLP models are also discussed in this chapter.

TensorFlow Steps to Build Models

In this section, I'll discuss the steps to build models in TensorFlow. I will walk you through the steps here, and then you'll see the code throughout this chapter:

1. Load the data.

2. Split the data into the train and test.

3. Normalize if needed.

4. Initialize placeholders that will contain predictors and the target.

© Navin Kumar Manaswi 2018
N. K. Manaswi, *Deep Learning with Applications Using Python*,
https://doi.org/10.1007/978-1-4842-3516-4_4

5. Create variables (weight and bias) that will be tuned up.

6. Declare model operations.

7. Declare the loss function and optimizer.

8. Initialize the variables and session.

9. Fit the model by using training loops.

10. Check and display the result with test data.

Linear Regression in TensorFlow

First you need to understand the code for linear regression in TensorFlow. Figure 4-1 shows a basic linear model.

As shown in Figure 4-1, the weight (W) and bias (b) are to be tuned so as to get the right values of weight and bias. So, the weight (W) and bias (b) are the variables in the TensorFlow code; you will tune/modify them in each iteration until you get the stable (correct) values.

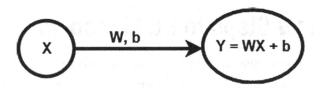

Figure 4-1. *Basic linear model*

You need to create placeholders for X. Placeholders have a particular shape and contain a particular type.

When you have more than one feature, you will have a similar working model to Figure 4-2.

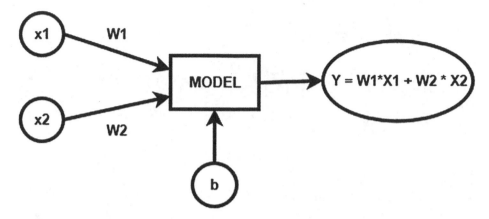

Figure 4-2. *Linear model with multiple inputs*

In the following code, you will use the Iris data set from Seaborn, which has five attributes. You will consider the sepal length as your input and the petal length as the output value. The main aim of this regression model is to predict the petal length when you are given the sepal length value. X is the sepal length, and Y is the petal length.

A linear regression using TensorFlow on Iris data

```
############ Linear Regression: TensorFlow Way ################
import matplotlib.pyplot as plt
import tensorflow as tf
from sklearn import datasets
import numpy as np
from sklearn.cross_validation import train_test_split
from matplotlib import pyplot
```

```
# 1. Load the data
# iris.data = [(Sepal Length, Sepal Width, Petal Length, Petal Width)]
iris = datasets.load_iris()
# X is Sepal.Length and Y is Petal Length
predictors_vals = np.array([predictors[0] for predictors in iris.data])
target_vals = np.array([predictors[2] for predictors in iris.data])
```

```
# 2.Split Data into train and test 80%-20%
x_trn, x_tst, y_trn, y_tst = train_test_split(predictors_vals, target_vals, test_size=0.2, random_state=12)
#training_idx = np.random.randint(x_vals.shape[0], size=80)
#training, test = x_vals[training_idx,:], x_vals[-training_idx,:]
```

```
# 3. Normalize if needed
# 4. Initialize placeholders that will contain predictors and target
predictor = tf.placeholder(shape=[None, 1], dtype=tf.float32)
target = tf.placeholder(shape=[None, 1], dtype=tf.float32)
```

```
#5. Create variables (Weight and Bias) that will be tuned up
A = tf.Variable(tf.zeros(shape=[1,1]))
b = tf.Variable(tf.ones(shape=[1,1]))
```

```
# 6. Declare model operations: Ax+b
model_output = tf.add(tf.matmul(predictor, A), b)
```

```
#7. Declare loss function and optimizer
#Declare loss function (L1 loss)
loss = tf.reduce_mean(tf.abs(target - model_output))
# Declare optimizer
my_opt = tf.train.GradientDescentOptimizer(0.01)
#my_opt = tftrain.AdamOptimizer(0.01)
train_step = my_opt.minimize(loss)
```

```
#8. Initialize variables and session
sess = tf.Session()
init = tf.global_variables_initializer()
sess.run(init)
```

```
#9. Fit Model by using Training Loops
# Training Loop
lossArray = []
batch_size = 40
for i in range(200):
    rand_rows = np.random.randint(0, len(x_trn)-1, size=batch_size)
    batchX = np.transpose([x_trn[rand_rows]])
    batchY = np.transpose([y_trn[rand_rows]])
    sess.run(train_step, feed_dict={predictor: batchX, target: batchY})
    batchLoss = sess.run(loss, feed_dict={predictor: batchX, target: batchY})
    lossArray.append(batchLoss)
    if (i+1)%50==0:
        print('Step Number' + str(i+1) + ' A = ' + str(sess.run(A)) + ' b = ' + str(sess.run(b)))
        print('L1 Loss = ' + str(batchLoss))

[slope] = sess.run(A)
[y_intercept] = sess.run(b)
```

```
# 10. Check  and Display the result on test data
lossArray = []
batch_size = 30
for i in range(100):
    rand_rows = np.random.randint(0, len(x_tst)-1, size=batch_size)
    batchX = np.transpose([x_tst[rand_rows]])
    batchY = np.transpose([y_tst[rand_rows]])
    sess.run(train_step, feed_dict={predictor: batchX, target: batchY})
    batchLoss = sess.run(loss, feed_dict={predictor: batchX, target: batchY})
    lossArray.append(batchLoss)
    if (i+1)%20==0:
        print('Step Number: ' + str(i+1) + ' A = ' + str(sess.run(A)) + ' b = ' + str(sess.run(b)))
        print('L1 Loss = ' + str(batchLoss))
# Get the optimal coefficients
[slope] = sess.run(A)
[y_intercept] = sess.run(b)
```

```
# Original Data and Plot
plt.plot(x_tst, y_tst, 'o', label='Actual Data')
test_fit = []
for i in x_tst:
  test_fit.append(slope*i+y_intercept)
# predicted values and Plot
plt.plot(x_tst, test_fit, 'r-', label='Predicted line', linewidth=3)
plt.legend(loc='lower right')
plt.title('Petal Length vs Sepal Length')
plt.ylabel('Petal Length')
plt.xlabel('Sepal Length')
plt.show()
```

```
# Plot Loss over time
plt.plot(lossArray, 'r-')
plt.title('L1 Loss per loop')
plt.xlabel('Loop')
plt.ylabel('L1 Loss')
plt.show()
```

If you run the code, you will see the output shown in Figure 4-3.

```
Step Number50 A = [[ 0.57060003]] b = [[ 1.05275011]]
L1 Loss = 0.965698
Step Number100 A = [[ 0.56647497]] b = [[ 1.00924981]]
L1 Loss = 1.124
Step Number150 A = [[ 0.56645012]] b = [[ 0.96424991]]
L1 Loss = 1.18043
Step Number200 A = [[ 0.58122498]] b = [[ 0.92174983]]
L1 Loss = 1.20376
Step Number: 20 A = [[ 0.58945829]] b = [[ 0.90308326]]
L1 Loss = 1.18207
Step Number: 40 A = [[ 0.62599164]] b = [[ 0.88975]]
L1 Loss = 0.826957
Step Number: 60 A = [[ 0.63695836]] b = [[ 0.87108338]]
L1 Loss = 0.838114
Step Number: 80 A = [[ 0.60072505]] b = [[ 0.8450833]]
L1 Loss = 1.52654
Step Number: 100 A = [[ 0.6150251]] b = [[ 0.8290832]]
L1 Loss = 1.25477
```

Figure 4-3. *Weights, bias, and loss at each step*

Figure 4-4 shows the plot for the predicted value of petal length.

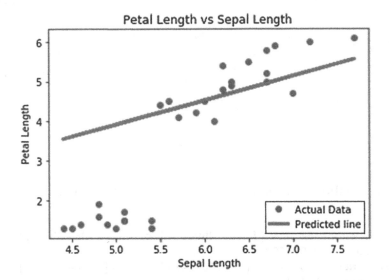

Figure 4-4. *Petal length versus sepal length*

Logistic Regression Model

For classification, the simplest approach is logistic regression. In this section, you'll learn how you can perform logistic regression in TensorFlow. Here you create the weight and bias as variables so that there is a scope of tuning/changing them per iteration. Placeholders are created to contain X. You need to create placeholders for X. Placeholders have a particular shape and contain a particular type, as shown in Figure 4-5.

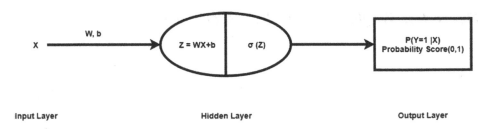

Figure 4-5. *Chart of logistic regression model*

In the following code, you will use the Iris data set, which has five attributes. The fifth one is the target class. You will consider the sepal length and sepal width as the predictor attributes and the flower's species as the target value. The main aim of this logistic regression model is to predict the kind of species when you are given the sepal length and sepal width values.

Create a Python file and import all the required libraries.

```python
######Logistic Regression in Tensorflow#########################
import numpy as np
import tensorflow as tf
from sklearn import datasets
import pandas as pd
from sklearn.cross_validation import train_test_split
from matplotlib import pyplot
# 1. Loading Data
iris = datasets.load_iris()
# Predictors Two columns : Sepal Length and Sepal Width
predictors_vals = np.array([predictor[0:2] for predictor in iris.data])
# For setosa Species, target is 0.
target_vals = np.array([1. if predictor==0 else 0. for predictor in iris.target])
```

```python
# 2. Split data into train/test - 75%/25%
predictors_vals_train, predictors_vals_test, target_vals_train, target_vals_test = train_test_split(predictors_vals, target_vals,
                                    train_size=0.75,
                                    random_state=0)
```

```python
# 3. Normalize if needed
#4.Initialize placeholders that will contain predictors and target
x_data = tf.placeholder(shape=[None, 2], dtype=tf.float32)
y_target = tf.placeholder(shape=[None, 1], dtype=tf.float32)
```

```python
#5. Create variables (Weight and Bias) that will be tuned up
W = tf.Variable(tf.ones(shape=[2,1]))
b = tf.Variable(tf.ones(shape=[1,1]))
```

```python
# 6. Declare model operations : y = xW +b
model = tf.add(tf.matmul(x_data, W), b)
```

```python
#7. Declare loss function and Optimizer
loss = tf.reduce_mean(tf.nn.sigmoid_cross_entropy_with_logits(logits=model, labels=y_target))
my_opt = tf.train.AdamOptimizer(0.02) #learning rate =0.02
train_step = my_opt.minimize(loss)
```

```python
#8. Initialize variables and session
init = tf.global_variables_initializer()
sess=tf.Session()
sess.run(init)
```

```
#9. Actual Prediction:
prediction = tf.round(tf.sigmoid(model))
predictions_correct = tf.cast(tf.equal(prediction, y_target), tf.float32)
accuracy = tf.reduce_mean(predictions_correct)
```

```
#10. Training loop
lossArray = []
trainAccuracy = []
testAccuracy = []
for i in range(1000):
    #Random instances for Batch size
    batch_size = 4 #Declare batch size
    batchIndex = np.random.choice(len(predictors_vals_train), size=batch_size)
    batchX = predictors_vals_train[batchIndex]
    batchY = np.transpose([target_vals_train[batchIndex]])
    # Tuning weight and bias while minimizing loss function through optimizer
    sess.run(train_step, feed_dict={x_data: batchX, y_target: batchY})
    #Loss function per epoch/generation
    batchLoss = sess.run(loss, feed_dict={x_data: batchX, y_target: batchY})
    lossArray.append(batchLoss) # adding it to loss_vec
    # accuracy for each epoch for train
    batchAccuracyTrain = sess.run(accuracy, feed_dict={x_data: predictors_vals_train, y_target: np.transpose([target_vals_trai
n])})
    trainAccuracy.append(batchAccuracyTrain) # adding it to train_acc
    # accuracy for each epoch for test
    batchAccuracyTest = sess.run(accuracy, feed_dict={x_data: predictors_vals_test, y_target: np.transpose([target_vals_test])})
    testAccuracy.append(batchAccuracyTest)
    # printing loss after 10 epochs/generations to avoid verbosity
    if (i+1)%50==0:
        print('Loss = ' + str(batchLoss)+ ' and Accuracy = '   + str(batchAccuracyTrain))
```

```
# 11. Check model performance
pyplot.plot(lossArray, 'r-' )
pyplot.title('Logistic Regression: Cross Entropy Loss per Epoch')
pyplot.xlabel('Epoch')
pyplot.ylabel('Cross Entropy Loss')
pyplot.show()
```

If you run the previous code, the plot of cross entropy loss against each epoch looks like Figure 4-6.

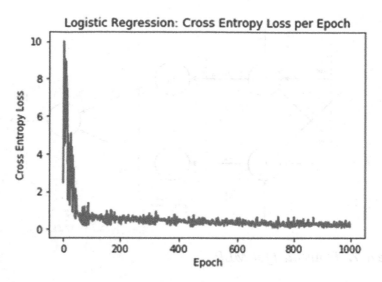

Figure 4-6. *Plot for cross entropy loss per epoch*

Multilayer Perceptron in TensorFlow

A multilayer perceptron (MLP) is a simple example of feedback artificial neural networks. An MLP consists of at least one hidden layer of nodes other than input layer and output layer. Each node of a layer other than the input layer is called a neuron that uses a nonlinear activation function such as sigmoid or ReLU. MLP uses a supervised learning technique called backpropagation for training while minimizing the loss function such as cross entropy and using an optimizer for tuning parameters (weight and bias). Its multiple layers and non-linear activation distinguish MLP from a linear perceptron.

TensorFlow is well suited for building MLP models. In an MLP, you need to tune the weight and bias per iteration. This means the weight and bias keep changing until they become stable while minimizing the loss function. So, you can create the weight and bias as variables in TensorFlow. I tend to give them initial values (all 0s or all 1s or some random normally distributed values). Placeholders should have a particular type of value and a defined shape, as shown in Figure 4-7.

65

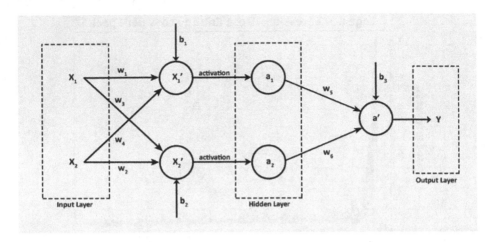

Figure 4-7. *Flowchart for MLP*

Import all the required libraries. Implementing MLP in TensorFlow.

```
# Implementing a one-hidden Layer Neural Network (MLP)

import matplotlib.pyplot as plt
import numpy as np
import tensorflow as tf
from sklearn import datasets
from sklearn.cross_validation import train_test_split
from matplotlib import pyplot
```

```
#2. Split data into train/test = 80%/20%

predictors_vals_train, predictors_vals_test, target_vals_train, target_vals_test= train_test_split(predictors_vals, target_val
s, test_size=0.2, random_state=12)
# 3. Normalize if needed
```

```
# 4.Initialize placeholders that will contain predictors and target
x_data = tf.placeholder(shape=[None, 3], dtype=tf.float32)
y_target = tf.placeholder(shape=[None, 1], dtype=tf.float32)
```

```
#5. Create variables (Weight and Bias) that will be tuned up
hidden_layer_nodes = 10
#For first Layer
A1 = tf.Variable(tf.ones(shape=[3,hidden_layer_nodes])) # inputs -> hidden nodes
b1 = tf.Variable(tf.ones(shape=[hidden_layer_nodes]))   # one biases for each hidden node
#For second layer
A2 = tf.Variable(tf.ones(shape=[hidden_layer_nodes,1])) # hidden inputs -> 1 output
b2 = tf.Variable(tf.ones(shape=[1]))    # 1 bias for the output
```

```
# 6. Define Model Structure
hidden_output = tf.nn.relu(tf.add(tf.matmul(x_data, A1), b1)) .
final_output = tf.nn.relu(tf.add(tf.matmul(hidden_output, A2), b2))
```

```
# 7. Declare loss function (MSE) and optimizer
loss = tf.reduce_mean(tf.square(y_target - final_output))
my_opt = tf.train.AdamOptimizer(0.02) # Learning rate = 0.02
train_step = my_opt.minimize(loss)
```

```
# 8.Initialize variables and session
init = tf.global_variables_initializer()
# Create graph session
sess = tf.Session()
sess.run(init)
```

```
# 9. Training Loop
lossArray = []
test_loss = []
batch_size =20
for i in range(500):
    batchIndex = np.random.choice(len(predictors_vals_train), size=batch_size)
    batchX = predictors_vals_train[batchIndex]
    batchY = np.transpose([target_vals_train[batchIndex]])
    sess.run(train_step, feed_dict={x_data: batchX, y_target: batchY})
    #
    batchLoss = sess.run(loss, feed_dict={x_data: batchX, y_target: batchY})
    lossArray.append(np.sqrt(batchLoss))

    test_temp_loss = sess.run(loss, feed_dict={x_data: predictors_vals_test, y_target: np.transpose([target_vals_test])})
    test_loss.append(np.sqrt(test_temp_loss))
    if (i+1)%50==0:
        print('Loss = ' + str(batchLoss))
```

```
# 10. Check model performance
#Plot Loss(mean squared error) over time
pyplot.plot(lossArray, 'o-', label='Train Loss')
pyplot.plot(test_loss, 'r--', label='Test Loss')
pyplot.title('Loss per Generation')
pyplot.legend(loc='lower left')
pyplot.xlabel('Generation')
pyplot.ylabel('Loss')
pyplot.show()
```

If you will run this code, you'll get the plot shown in Figure 4-8.

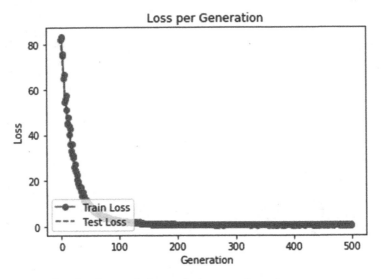

Figure 4-8. *Plot for loss while training and testing*

In this chapter, I discussed how you can build linear, logistic, and MLP models in TensorFlow in a systemic way.

## CHAPTER 5

Regression to MLP in Keras

You have been working on regression while solving machine learning applications. Linear regression and nonlinear regression are used to predict numeric targets, while logistic regression and other classifiers are used to predict non-numeric target variables. In this chapter, I will discuss the evolution of multilayer perceptrons.

Specifically, you will compare the accuracy generated by different models with and without using Keras.

Log-Linear Model

Create a new Python file and import the following packages. Make sure you have Keras installed on your system.

```
###### Log-Linear Model ######################################
from sklearn.datasets import load_iris
from sklearn.cross_validation import train_test_split
from sklearn.linear_model import LogisticRegressionCV
from sklearn.linear_model import LinearRegression
import numpy as np
import matplotlib.pyplot as plt
from keras.models import Sequential
from keras.layers import Dense, Activation
```

© Navin Kumar Manaswi 2018
N. K. Manaswi, *Deep Learning with Applications Using Python*,
https://doi.org/10.1007/978-1-4842-3516-4_5

You will be using the Iris data set as the source of data. So, load the data set from Seaborn.

```
# Load the iris dataset from seaborn.
iris = load_iris()
```

The Iris data set has five attributes. You will be using the first four attributes to predict the species, whose class is defined in the fifth attribute of the data set.

```
# Use the first 4 variables to predict the species.
X, y = iris.data[:, :4], iris.target
```

Using `scikit-learn`'s function, split the testing and training data sets.

```
# Split both independent and dependent variables in half
# for cross-validation
train_X, test_X, train_y, test_y = train_test_split(X, y, train_size=0.5, random_state=0)
#print(type(train_X),len(train_y),len(test_X),len(test_y))
```

```
####################################
# scikit Learn for (Log) Linear Regression #
####################################
```

Use the `model.fit` function to train the model with the training data set.

```
# Train a scikit-Learn Log-regression model
# Lr =LogisticRegressionCV
# Train a scikit-Learn linear-regression model
lr = LinearRegression()
lr.fit(train_X, train_y)
```

As the model is trained, you can predict the output of the test set.

```
# Test the model. Print the accuracy on the test data
pred_y = lr.predict(test_X)
#print("Accuracy is {:.2f}".format(lr.score(test_X, test_y)))
```

Keras Neural Network for Linear Regression

Now, let's build a Keras neural network model for linear regression.

```
# Build the keras model
model = Sequential()
# 4 features in the input layer (the four flower measurements)
# 16 hidden units
model.add(Dense(16, input_shape=(4,)))
model.add(Activation('sigmoid'))
# 3 classes in the ouput layer (corresponding to the 3 species)
model.add(Dense(3))
model.add(Activation('softmax'))
```

```
# Compile the model
model.compile(loss='sparse_categorical_crossentropy', optimizer='adam', metrics=['accuracy'])
```

Use the `model.fit` function to train the model with the training data set.

```
# Fit/Train the keras model
model.fit(train_X, train_y, verbose=1, batch_size=1, nb_epoch=100)
```

As the model is trained, you can predict the output of the test set.

```
# Test the model. Print the accuracy on the test data
loss, accuracy = model.evaluate(test_X, test_y, verbose=0)
print("\nAccuracy is using keras prediction  {:.2f}".format(accuracy))
```

Print the accuracy obtained by both models.

```
print("\nAccuracy is using keras prediction  {:.2f}".format(accuracy))

print("Accuracy is using regression  {:.2f}".format(lr.score(test_X, test_y)))
```

If you run the code, you will see the following output:

```
Using TensorFlow backend.
Epoch 1/100
75/75 [==============================] - 0s - loss: 1.2947 - acc: 0.4533
Epoch 2/100
75/75 [==============================] - 0s - loss: 1.0353 - acc: 0.6400
Epoch 3/100
75/75 [==============================] - 0s - loss: 0.8930 - acc: 0.6533
...                    ...                    ...

...                    ...                    ...

...                    ...                    ...

...                    ...                    ...

Epoch 99/100
75/75 [==============================] - 0s - loss: 0.1186 - acc: 0.9733
Epoch 100/100
75/75 [==============================] - 0s - loss: 0.1167 - acc: 0.9867


Accuracy is using keras prediction  0.99
Accuracy is using regression  0.89
```

Logistic Regression

In this section, I will share an example for the logistic regression so you can compare the code in `scikit-learn` with that in Keras (see Figure 5-1).

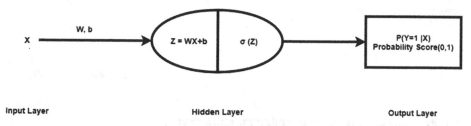

Figure 5-1. *Logistic regression used for classification*

Create a new Python file and import the following packages. Make sure you have Keras installed on your system.

```
from sklearn.datasets import load_iris
import numpy as np
from sklearn.cross_validation import train_test_split
from sklearn.linear_model import LogisticRegressionCV
from keras.models import Sequential
from keras.layers.core import Dense, Activation
from keras.utils import np_utils
```

You will be using the Iris data set as the source of data. So, load the data set from scikit-learn.

```
# Load Data and Prepare data
iris = load_iris()
X, y = iris.data[:, :4], iris.target
```

Using `scikit-learn`'s function, split the testing and training data sets.

```
# Load Data and Prepare data
iris = load_iris()
X, y = iris.data[:, :4], iris.target
```

scikit-learn for Logistic Regression

Use the `model.fit` function to train the model with the training data set. After the model is trained, you can predict the output of the test set.

```
###############################
# scikit Learn for Logistic Regression
###############################
lr = LogisticRegressionCV()
lr.fit(train_X, train_y)
pred_y = lr.predict(test_X)
print("Test fraction correct (LR-Accuracy) = {:.2f}".format(lr.score(test_X, test_y)))
```

##

Keras Neural Network for Logistic Regression

One-hot encoding transforms features to a format that works better with the classification and regression algorithms.

```
###############################
# Keras Neural Network for Logistic Regression
###############################

# Make ONE-HOT enconding for converting into categorical variable
def one_hot_encode_object_array(arr):
    uniques, ids = np.unique(arr, return_inverse=True)
    return np_utils.to_categorical(ids, len(uniques))
```

```
# Dividing data into train and test data
train_y_ohe = one_hot_encode_object_array(train_y)
test_y_ohe = one_hot_encode_object_array(test_y)
#Creating a model
model = Sequential()
model.add(Dense(16, input_shape=(4,)))
model.add(Activation('sigmoid'))
model.add(Dense(3))
model.add(Activation('softmax'))
```

```
# Compiling the model
model.compile(loss='categorical_crossentropy', metrics=['accuracy'], optimizer='adam')
```

Use the model.fit function to train the model with the training data set.

```
# Actual modelling
model.fit(train_X, train_y_ohe, verbose=0, batch_size=1, nb_epoch=100)
```

Use the model.evaluate function to evaluate the performance of the model.

```
score, accuracy = model.evaluate(test_X, test_y_ohe, batch_size=16, verbose=0)
```

Print the accuracy obtained by both models.

Accuracy for scikit-learn based model

```
print("\n Test fraction correct (LR-Accuracy) logistic regression = {:.2f}".format(lr.score(test_X, test_y)))
```

The accuracy is 0.83.

Accuracy for keras model

```
print("Test fraction correct (NN-Accuracy) keras   = {:.2f}".format(accuracy))
```

The accuracy is 0.99.

If you run the code, you will see the following output:

```
Using TensorFlow backend.
Test fraction correct (LR-Accuracy) logistic regression =
0.83
Test fraction correct (NN-Accuracy) keras  = 0.99
Epoch 1/100
75/75 [==============================] - 0s - loss: 1.2947 -
acc: 0.4533
Epoch 2/100
75/75 [==============================] - 0s - loss: 1.0353 -
acc: 0.6400
Epoch 3/100
75/75 [==============================] - 0s - loss: 0.8930 -
acc: 0.6533
...                                    ...                ...
...                                    ...                ...
...                                    ...                ...
...                                    ...                ...
Epoch 99/100
75/75 [==============================] - 0s - loss: 0.1186 -
acc: 0.9733
Epoch 100/100
75/75 [==============================] - 0s - loss: 0.1167 -
acc: 0.9867

Accuracy is using keras prediction  0.99
Accuracy is using regression  0.89
```

To give the real-life example, I will discuss some code that uses the Fashion MNIST data set, which is a data set of Zalando.com's images consisting of a training set of 60,000 examples and a test set of 10,000 examples. Each example is a 28×28 grayscale image associated with a label from ten classes.

Fashion MNIST Data: Logistic Regression in Keras

Create a new Python file and import the following packages. Make sure you have Keras installed on your system.

```python
from __future__ import print_function
from keras.models import load_model
import keras
import fashion_mnist
from keras.models import Sequential
from keras.layers import Dense, Dropout
import numpy as np

batch_size = 128
num_classes = 10
epochs = 2
```

As mentioned, you will be using the Fashion MNIST data set. Store the data and the label in two different variables.

```python
# the data, shuffled and split between train and test sets
(x_train, y_train), (x_test, y_test) = fashion_mnist.load_data()

x_train = x_train.reshape(60000, 784)
x_test = x_test.reshape(10000, 784)
x_train = x_train.astype('float32')
x_test = x_test.astype('float32')
```

Normalize the data set, as shown here:

```python
#Gaussian Normalization of the dataset
x_train = (x_train-np.mean(x_train))/np.std(x_train)
x_test = (x_test-np.mean(x_test))/np.std(x_test)

# convert class vectors to binary class matrices
y_train = keras.utils.to_categorical(y_train, num_classes)
y_test = keras.utils.to_categorical(y_test, num_classes)
```

Define the model, as shown here:

```
#Building a model architecture
model = Sequential()
model.add(Dense(256, activation='elu', input_shape=(784,)))
model.add(Dropout(0.4))
model.add(Dense(512, activation='relu'))
model.add(Dense(num_classes, activation='softmax'))

model.summary()

model.compile(loss='categorical_crossentropy',
              optimizer='adam',
              metrics=['accuracy'])

model.fit(x_train, y_train,
                  batch_size=batch_size,
                  epochs=epochs,
                  validation_data=(x_test, y_test))
```

Save the model in an .h5 file (so that you can use it later directly with the load_model() function from keras.models) and print the accuracy of the model in the test set, as shown here:

```
#saving the model using the 'model.save' function
model.save('my_model.h5')
score = model.evaluate(x_test, y_test, verbose=0)
print('Test loss:', score[0])
print('Test accuracy:', score[1])
```

If you run the previous code, you will see the following output:

```
('train-images-idx3-ubyte.gz', <http.client.HTTPMessage object
at 0x00000171338E2B38>)
```

Layer (type)	Output Shape	Param #
dense_59 (Dense)	(None, 256)	200960
dropout_10 (Dropout)	(None, 256)	0
dense_60 (Dense)	(None, 512)	131584
dense_61 (Dense)	(None, 10)	5130

```
Total params: 337,674
Trainable params: 337,674
Non-trainable params: 0
```

```
Train on 60000 samples, validate on 10000 samples
Epoch 1/2
60000/60000 [==============================] - loss: 0.5188 -
acc: 0.8127 - val_loss: 0.4133 - val_acc: 0.8454
Epoch 2/2
60000/60000 [==============================] - loss: 0.3976 -
acc: 0.8545 - val_loss: 0.4010 - val_acc: 0.8513
Test loss: 0.400989927697
Test accuracy: 0.8513
```

MLPs on the Iris Data

A multilayer perceptron is a minimal neural network model. In this section, I'll show you the code.

Write the Code

Create a new Python file and import the following packages. Make sure you have Keras installed on your system.

```
###########MLP on iris data ####################

import pandas as pd
import numpy as np
from keras.models import Sequential
from keras.layers import Dense, Activation
from keras.utils import np_utils
```

Load the data set by reading a CSV file using Pandas.

```
#Load and Prepare Data
datatrain = pd.read_csv('./Datasets/iris/iris_train.csv')
```

Assign numeric values to the classes of the data set.

```
#change string value to numeric
datatrain.set_value(datatrain['species']=='Iris-setosa',['species'],0)
datatrain.set_value(datatrain['species']=='Iris-versicolor',['species'],1)
datatrain.set_value(datatrain['species']=='Iris-virginica',['species'],2)
datatrain = datatrain.apply(pd.to_numeric)
```

Convert the data frame to an array.

```
#change dataframe to array
datatrain_array = datatrain.as_matrix()
```

Split the data and the target and store them in two different variables.

```
# split x and y (feature and target)
xtrain = datatrain_array[:,:4]
ytrain = datatrain_array[:,4]
```

Change the target format using Numpy.

```
#change target format
ytrain = np_utils.to_categorical(ytrain)
```

Build a Sequential Keras Model

Here you will build a multilayer perceptron model with one hidden layer.

- *Input layer*: The input layer contains four neurons, representing the features of an iris (sepal length, etc.).

- *Hidden layer*: The hidden layer contains ten neurons, and the activation uses ReLU.

- *Output layer*: The output layer contains three neurons, representing the classes of the Iris softmax layer.

```
#Build Keras model
#Multilayer perceptron model, with one hidden layer.
#Input layer : 4 neuron, represents the feature of Iris(Sepal Length etc)
#Hidden Layer : 10 neuron, activation using ReLU
#Output Layer : 3 neuron, represents the class of Iris, Softmax Layer
model = Sequential()
model.add(Dense(output_dim=10, input_dim=4))
model.add(Activation("relu"))
model.add(Dense(output_dim=3))
model.add(Activation("softmax"))
```

Compile the model and choose an optimizer and loss function for training and optimizing your data, as shown here:

```
#Compile model :choose optimizer and loss function
#optimizer = stochastic gradient descent with no batch-size
#loss function = categorical cross entropy
#learning rate = default from keras.optimizer.SGD, 0.01
model.compile(loss='categorical_crossentropy', optimizer='sgd', metrics=['accuracy'])
```

Train the model using the `model.fit` function, as shown here:

```
#train
model.fit(xtrain, ytrain, nb_epoch=100, batch_size=120)
```

Load and prepare the test data, as shown here:

```
## Evaluate on test data
#Load and Prepare Data
datatest = pd.read_csv('./Datasets/iris/iris_test.csv')
```

Convert the string value to a numeric value, as shown here:

```
#change string value to numeric
datatest.set_value(datatest['species']=='Iris-setosa',['species'],0)
datatest.set_value(datatest['species']=='Iris-versicolor',['species'],1)
datatest.set_value(datatest['species']=='Iris-virginica',['species'],2)
datatest = datatest.apply(pd.to_numeric)
```

Convert the data frame to an array, as shown here:

```
#change dataframe to array
datatest_array = datatest.as_matrix()
```

Split x and y, in other words, the feature set and target set, as shown here:

```
#split x and y (feature and target)
xtest= datatest_array[:,:4]
ytest = datatest_array[:,4]
```

Make a prediction on the trained model, as shown here:

```
#get prediction
classes = model.predict_classes(xtest, batch_size=120)
```

Calculate the accuracy, as shown here:

```
#get accuration
accuration = np.sum(classes == ytest)/30.0 * 100
```

Print the accuracy generated by the model, as shown here:

```
print("Test Accuration : " + str(accuration) + '%')
print("Prediction :")
print(classes)
print("Target :")
print(np.asarray(ytest,dtype="int32"))
```

If you run the code, you will see the following output:

```
Epoch 1/100
120/120 [==============================] - 0s - loss: 2.7240 -
acc: 0.3667
Epoch 2/100
120/120 [==============================] - 0s - loss: 2.4166 -
acc: 0.3667
Epoch 3/100
120/120 [==============================] - 0s - loss: 2.1622 -
acc: 0.4083
Epoch 4/100
120/120 [==============================] - 0s - loss: 1.9456 -
acc: 0.6583
```

```
Epoch 98/100
120/120 [==============================] - 0s - loss: 0.5571 -
acc: 0.9250
Epoch 99/100
120/120 [==============================] - 0s - loss: 0.5554 -
acc: 0.9250
Epoch 100/100
120/120 [==============================] - 0s - loss: 0.5537 -
acc: 0.9250
```

MLPs on MNIST Data (Digit Classification)

MNIST is the standard data set to predict handwritten digits. In this section, you will see how you can apply the concept of multilayer perceptrons and make a handwritten digit recognition system.

Create a new Python file and import the following packages. Make sure you have Keras installed on your system.

```
#########MLP : MNIST Data (Digit Classification) #######################

import numpy as np
import os
from keras.datasets import mnist
from keras.models import Sequential
from keras.layers.core import Dense, Dropout, Activation
from keras.optimizers import RMSprop
from keras.utils import np_utils
```

Sone important variables are defined.

```
np.random.seed(100) # for reproducibility
batch_size = 128 #Number of images used in each optimization step
nb_classes = 10 #One class per digit
nb_epoch = 20 #Number of times the whole data is used to learn
```

Load the data set using the `mnist.load_data()` function.

```
(X_train, y_train), (X_test, y_test) = mnist.load_data()

#Flatten the data, MLP doesn't use the 2D structure of the data. 784 = 28*28
X_train = X_train.reshape(60000, 784) # 60,000 digit images
X_test = X_test.reshape(10000, 784)
```

The types of the training set and the test set are converted to `float32`.

```
X_train = X_train.astype('float32')
X_test = X_test.astype('float32')
```

The data sets are normalized; in other words, they are set to a Z-score.

```
# Gaussian Normalization( Z- score)
X_train = (X_train- np.mean(X_train))/np.std(X_train)
X_test = (X_test- np.mean(X_test))/np.std(X_test)
```

Display the number of the training samples present in the data set and also the number of test sets available.

```
#Display number of training and test instances
print(X_train.shape[0], 'train samples')
print(X_test.shape[0], 'test samples')
```

Convert class vectors to binary class matrices.

```
# convert class vectors to binary class matrices (ie one-hot vectors)
Y_train = np_utils.to_categorical(y_train, nb_classes)
Y_test = np_utils.to_categorical(y_test, nb_classes)
```

Define the sequential model of the multilayer perceptron.

```
#Define the model achitecture
model = Sequential()
model.add(Dense(512, input_shape=(784,)))
model.add(Activation('relu'))
model.add(Dropout(0.2)) # Regularization
model.add(Dense(120))
model.add(Activation('relu'))
model.add(Dropout(0.2))
model.add(Dense(10)) #Last layer with one output per class
model.add(Activation('softmax')) #We want a score simlar to a probability for each class
```

Use an optimizer.

```
#Use rmsprop as an optimizer
rms = RMSprop()
```

The function to optimize is the cross entropy between the true label and the output (softmax) of the model.

```
#The function to optimize is the cross entropy between the true label and the output (softmax) of the model
model.compile(loss='categorical_crossentropy', optimizer=rms, metrics=["accuracy"])
```

Use the `model.fit` function to train the model.

```
#Make the model learn
model.fit(X_train, Y_train,
batch_size=batch_size, nb_epoch=nb_epoch,
verbose=2,
validation_data=(X_test, Y_test))
```

Using the model, evaluate the function to evaluate the performance of the model.

```
#Evaluate how the model does on the test set
score = model.evaluate(X_test, Y_test, verbose=0)
```

Print the accuracy generated in the model.

```
print('Test score:', score[0])
print('Test accuracy:', score[1])
```

If you run the code, you will get the following output:

60000 train samples

```
10000 test samples
Train on 60000 samples, validate on 10000 samples
Epoch 1/20
13s - loss: 0.2849 - acc: 0.9132 - val_loss: 0.1149 - val_acc:
0.9652
Epoch 2/20
11s - loss: 0.1299 - acc: 0.9611 - val_loss: 0.0880 - val_acc:
0.9741
Epoch 3/20
11s - loss: 0.0998 - acc: 0.9712 - val_loss: 0.1121 - val_acc:
0.9671
Epoch 4/20
Epoch 18/20
14s - loss: 0.0538 - acc: 0.9886 - val_loss: 0.1241 - val_acc:
0.9814
Epoch 19/20
12s - loss: 0.0522 - acc: 0.9888 - val_loss: 0.1154 - val_acc:
0.9829
Epoch 20/20
13s - loss: 0.0521 - acc: 0.9891 - val_loss: 0.1183 - val_acc:
0.9824
Test score: 0.118255248802
Test accuracy: 0.9824
```

Now, it is time to create a data set and use a multilayer perceptron. Here you will create your own data set using the random function and run the multilayer perceptron model on the generated data.

MLPs on Randomly Generated Data

Create a new Python file and import the following packages. Make sure you have Keras installed on your system.

```
#####MLP on randomly generated Data #############
import keras
from keras.models import Sequential
from keras.layers import Dense, Dropout, Activation
from keras.optimizers import SGD
import numpy as np
```

Generate the data using the random function.

```
# Generate dummy data
x_train = np.random.random((1000, 20))
# Y having 10 possible categories
y_train = keras.utils.to_categorical(np.random.randint(10, size=(1000, 1)), num_classes=10)
x_test = np.random.random((100, 20))
y_test = keras.utils.to_categorical(np.random.randint(10, size=(100, 1)), num_classes=10)
```

Create a sequential model.

```
#Create a model
model = Sequential()
# Dense(64) is a fully-connected layer with 64 hidden units.
# In the first layer, you must specify the expected input data shape:
# here, 20-dimensional vectors.
model.add(Dense(64, activation='relu', input_dim=20))
model.add(Dropout(0.5))
model.add(Dense(64, activation='relu'))
model.add(Dropout(0.5))
model.add(Dense(10, activation='softmax'))
```

Compile the model.

```
#Compile the model
sgd = SGD(lr=0.01, decay=1e-6, momentum=0.9, nesterov=True)
model.compile(loss='categorical_crossentropy',optimizer=sgd,metrics=['accuracy'])
```

Use the model.fit function to train the model.

```
# Fit the model
model.fit(x_train, y_train,epochs=20,batch_size=128)
```

Evaluate the performance of the model using the model.evaluate function.

```
# Evaluate the model
score = model.evaluate(x_test, y_test, batch_size=128)
```

If you run the code, you will get the following output:

```
Epoch 1/20
1000/1000 [==============================] - 0s - loss:
2.4432 - acc: 0.0970
Epoch 2/20
1000/1000 [==============================] - 0s - loss:
2.3927 - acc: 0.0850
Epoch 3/20
1000/1000 [==============================] - 0s - loss:
2.3361 - acc: 0.1190
Epoch 4/20
1000/1000 [==============================] - 0s - loss:
2.3354 - acc: 0.1000
Epoch 19/20
1000/1000 [==============================] - 0s - loss:
2.3034 - acc: 0.1160
Epoch 20/20
1000/1000 [==============================] - 0s - loss:
2.3055 - acc: 0.0980
100/100 [==============================] - 0s
```

In this chapter, I discussed how to build linear, logistic, and MLP models in Keras in a systemic way.

CHAPTER 6

Convolutional Neural Networks

A *convolutional neural network* (CNN) is a deep, feed-forward artificial neural network in which the neural network preserves the hierarchical structure by learning internal feature representations and generalizing the features in the common image problems like object recognition and other computer vision problems. It is not restricted to images; it also achieves state-of-the-art results in natural language processing problems and speech recognition.

Different Layers in a CNN

A CNN consists of multiple layers, as shown in Figure 6-1.

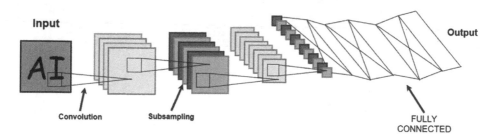

Figure 6-1. *Layers in a convolution neural network*

N. K. Manaswi, *Deep Learning with Applications Using Python*,
https://doi.org/10.1007/978-1-4842-3516-4_6

The *convolution layers* consist of filters and image maps. Consider the grayscale input image to have a size of 5×5, which is a matrix of 25 pixel values. The image data is expressed as a three-dimensional matrix of width × height × channels.

Note   An image map is a list of coordinates relating to a specific image.

Convolution aims to extract features from the input image, and hence it preserves the spatial relationship between pixels by learning image features using small squares of input data. Rotational invariance, translation invariance, and scale invariance can be expected. For example, a rotated cat image or rescaled cat image can be easily identified by a CNN because of the convolution step. You slide the filter (square matrix) over your original image (here, 1 pixel), and at each given position, you compute element-wise multiplication (between the matrices of the filter and the original image) and add the multiplication outputs to get the final integer that forms the elements of the output matrix.

Subsampling is simply the average pooling with learnable weights per feature map, as shown in Figure 6-2.

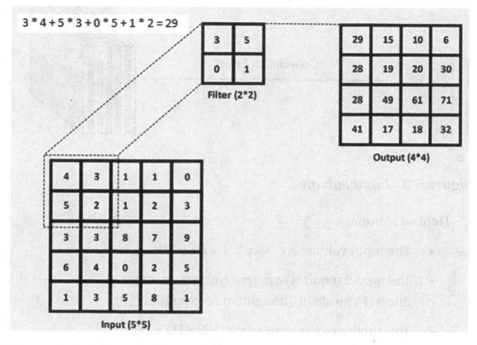

Figure 6-2. *Subsampling*

As shown in Figure 6-2, filters have input weights and generate an output neuron. Let's say you define a convolutional layer with six filters and receptive fields that are 2 pixels wide and 2 pixels high and use a default stride width of 1, and the default padding is set to 0. Each filter receives input from 2×2 pixels, section of image. In other words, that's 4 pixels at a time. Hence, you can say it will require 4 + 1 (bias) input weights.

The input volume is 5×5×3 (width × height × number of channel), there are six filters of size 2×2 with stride 1 and pad 0. Hence, the number of parameters in this layer for each filter has 2*2*3 + 1 = 13 parameters (added +1 for bias). Since there are six filters, you get 13*6 = 78 parameters.

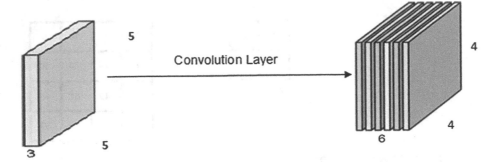

Figure 6-3. *Input volume*

Here's a summary:

- The input volume is of size W1 × H1 × D1.

- The model requires hyperparameters: number of filters (f), stride (S), amount of zero padding (P).

- This produces a volume of size W2 × H2 × D2.

- W2 = (W1-f+ 2P) /S + 1 = 4.

- H2 = (H1-f+2P)/S +1 = 4.

- D2 = Number of filters = f = 6.

The pooling layers reduce the previous layers' activation maps. It is followed by one or more convolutional layers and consolidates all the features that were learned in the previous layers' activation maps. This reduces the overfitting of the training data and generalizes the features represented by the network. The receptive field size is almost always set to 2×2 and use a stride of 1 or 2 (or higher) to ensure there is no overlap. You will use a max operation for each receptive field so that the activation is the maximum input value. Here, every four numbers map to just one number. So, the number of pixels goes down to one-fourth of the original in this step (Figure 6-4).

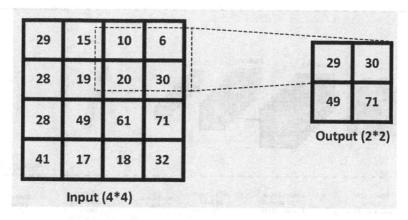

Figure 6-4. *Maxpooling-reducing the number of pixels*

A fully connected layer is a feed-forward artificial neural network layer. These layers have a nonlinear activation function to output class prediction probabilities. They are used toward the end after all the features are identified and extracted by convolutional layers and have been consolidated by the pooling layers in the network. Here, the hidden and output layers are the fully connected layers.

CNN Architectures

A CNN is a feed-forward deep neural network architecture comprised of a few convolutional layers, each followed by a pooling layer, activation function, and optionally batch normalization. It also comprises of the fully connected layers. As an image moves through the network, it gets smaller, mostly because of max pooling. The final layer outputs the class probabilities prediction.

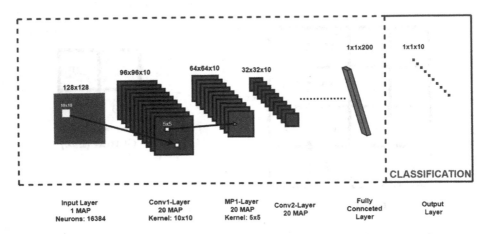

Figure 6-5. *CNN Architecture for Classification*

The past few years have seen many architectures being developed that have made tremendous progress in the field of image classification. Award-winning pretrained networks (VGG16, VGG19, ResNet50, Inception V3, and Xception) have been used for various image classification challenges including medical imaging. Transfer learning is the kind of practice where you use pretrained models in addition to a couple of layers. It can be used to solve image classification challenges in every field.

CHAPTER 7

CNN in TensorFlow

This chapter will demonstrate how to use TensorFlow to build a CNN model. A CNN model can help you build an image classifier that can predict/classify the images. In general, you create some layers in the model architecture with initial values of weight and bias. Then you tune weight and bias with the help of a training data set. There is another approach that involves using a pretrained model such as InceptionV3 to classify the images. You can use this transfer learning approach where you add some layers (whose parameter s are trained) on top of layers of pretrained models (with parameter values intact) to make very powerful classifiers.

In this chapter, I will use TensorFlow to show how to develop a convolution network for various computer vision applications. It is easier to express a CNN architecture as a graph of data flows.

Why TensorFlow for CNN Models?

In TensorFlow, images can be represented as three-dimensional arrays or tensors of shape (height, width and channels). TensorFlow provides the flexibility to quickly iterate, allows you to train models faster, and enables you to run more experiments. When taking TensorFlow models to production, you can run them on large-scale GPUs and TPUs.

© Navin Kumar Manaswi 2018
N. K. Manaswi, *Deep Learning with Applications Using Python*,
https://doi.org/10.1007/978-1-4842-3516-4_7

TensorFlow Code for Building an Image Classifier for MNIST Data

In this section, I'll take you through an example to understand how to implement a CNN in TensorFlow.

The following code imports MNIST data sets with 28×28 grayscale images of digits from the TensorFlow contrib package and loads all the required libraries. Here, the aim is to build the classifier to predict the digit given in the image.

```
from tensorflow.contrib.learn.python.learn.datasets.mnist
import read_data_sets
from tensorflow.python.framework import ops
import tensorflow as tf
import numpy as np
```

You then start a graph session.

```
# Start a graph session
sess = tf.Session()
```

You load the MNIST data and create the train and test sets.

```
# Load data
from keras.datasets import mnist
(X_train, y_train), (X_test, y_test) = mnist.load_data()
```

You then normalize the train and test set features.

```
# Z- score  or Gaussian Normalization
X_train = X_train - np.mean(X_train) / X_train.std()
X_test = X_test - np.mean(X_test) / X_test.std()
```

As this is a multiclass classification problem, it is always better to use the one-hot encoding of the output class values.

```
# Convert labels into one-hot encoded vectors
num_class = 10
train_labels = tf.one_hot(y_train, num_class)
test_labels = tf.one_hot(y_test, num_class)
```

Let's set the model parameters now as these images are grayscale. Hence, the depth of image (channel) is 1.

```
# Set model parameters
batch_size = 784
samples =500
learning_rate = 0.03
img_width = X_train[0].shape[0]
img_height = X_train[0].shape[1]
target_size = max(train_labels) + 1
num_channels = 1 # greyscale = 1 channel
epoch = 200
no_channels = 1
conv1_features = 30
filt1_features = 5
conv2_features = 15
filt2_features = 3
max_pool_size1 = 2 # NxN window for 1st max pool layer
max_pool_size2 = 2 # NxN window for 2nd max pool layer
fully_connected_size1 = 150
```

Let's declare the placeholders for the model. The input data features, target variable, and batch sizes can be changed for the training and evaluation sets.

```
# Declare model placeholders
x_input_shape = (batch_size, img_width, img_height, num_channels)
x_input = tf.placeholder(tf.float32, shape=x_input_shape)
y_target = tf.placeholder(tf.int32, shape=(batch_size))
eval_input_shape = (samples, img_width, img_height, num_channels)
eval_input = tf.placeholder(tf.float32, shape=eval_input_shape)
eval_target = tf.placeholder(tf.int32, shape=(samples))
```

Let's declare the model variables' weight and bias values for input and hidden layer's neurons.

```
# Declare model variables
W1 = tf.Variable(tf.random_normal([filt1_features,
filt1_features, no_channels, conv1_features]))
b1 = tf.Variable(tf.ones([conv1_features]))
W2 = tf.Variable(tf.random_normal([filt2_features,
filt2_features, conv1_features, conv2_features]))
b2 = tf.Variable(tf.ones([conv2_features]))
```

Let's declare the model variables for fully connected layers and define the weights and bias for these last 2 layers.

```
# Declare model variables for fully connected layers
resulting_width = img_width // (max_pool_size1 * max_pool_size2)
resulting_height = img_height // (max_pool_size1 * max_pool_size2)
full1_input_size = resulting_width * resulting_height * conv2_
features
W3 = tf.Variable(tf.truncated_normal([full1_input_size,
fully_connected_size1], stddev=0.1, dtype=tf.float32))
```

```
b3 = tf.Variable(tf.truncated_normal([fully_connected_size1],
stddev=0.1, dtype=tf.float32))
W_out = tf.Variable(tf.truncated_normal([fully_connected_size1,
target_size], stddev=0.1, dtype=tf.float32))
b_out = tf.Variable(tf.truncated_normal([target_size],
stddev=0.1, dtype=tf.float32))
```

Let's create a helper function to define the convolutional and max pooling layers.

```
# Define helper functions for the convolution and maxpool layers:
def conv_layer(x, W, b):
    conv = tf.nn.conv2d(x, W, strides=[1, 1, 1, 1],
    padding='SAME')
    conv_with_b = tf.nn.bias_add(conv, b)
    conv_out = tf.nn.relu(conv_with_b)
    return conv_out
def maxpool_layer(conv, k=2):
    return tf.nn.max_pool(conv, ksize=[1, k, k, 1],
    strides=[1, k, k, 1], padding='SAME')
```

A neural network model is defined with two hidden layers and two fully connected layers. A rectified linear unit is used as the activation function for the hidden layers and the final output layers.

```
# Initialize Model Operations
def my_conv_net(input_data):
    # First Conv-ReLU-MaxPool Layer
    conv_out1 = conv_layer(input_data, W1, b1)
    maxpool_out1 = maxpool_layer(conv_out1)
```

```
    # Second Conv-ReLU-MaxPool Layer
    conv_out2 = conv_layer(maxpool_out1, W2, b2)
    maxpool_out2 = maxpool_layer(conv_out2)

    # Transform Output into a 1xN layer for next fully
    connected layer
    final_conv_shape = maxpool_out2.get_shape().as_list()
    final_shape = final_conv_shape[1] * final_conv_shape[2] *
    final_conv_shape[3]
    flat_output = tf.reshape(maxpool_out2, [final_conv_shape[0],
    final_shape])

    # First Fully Connected Layer
    fully_connected1 = tf.nn.relu(tf.add(tf.matmul(flat_output,
    W3), b3))
    # Second Fully Connected Layer
    final_model_output = tf.add(tf.matmul(fully_connected1,
    W_out), b_out)

    return(final_model_output)
model_output = my_conv_net(x_input)
test_model_output = my_conv_net(eval_input)
```

You will use a softmax cross entropy function (tends to work better for multiclass classification) to define the loss that operates on logits.

```
# Declare Loss Function (softmax cross entropy)
loss = tf.reduce_mean(tf.nn.sparse_softmax_cross_entropy_with_
logits(logits=model_output, labels=y_target))
```

Let's define the train and test sets' prediction function.

```
# Create a prediction function
prediction = tf.nn.softmax(model_output)
test_prediction = tf.nn.softmax(test_model_output)
```

To determine the model accuracy on each batch, let's define the accuracy function.

```
# Create accuracy function
def get_accuracy(logits, targets):
    batch_predictions = np.argmax(logits, axis=1)
    num_correct = np.sum(np.equal(batch_predictions, targets))
    return(100. * num_correct/batch_predictions.shape[0])
```

Let's declare the training step and define the optimizer function.

```
# Create an optimizer
my_optimizer = tf.train.AdamOptimizer(learning_rate, 0.9)
train_step = my_optimizer.minimize(loss)
```

Let's initialize all the model variables declared earlier.

```
# Initialize Variables
varInit = tf.global_variables_initializer()
sess.run(varInit)
```

Let's start training the model and loop randomly through the batches of data. You want to evaluate the model on the train and test set batches and record the loss and accuracy.

```
# Start training loop
train_loss = []
train_acc = []
test_acc = []
for i in range(epoch):
    random_index = np.random.choice(len(X_train), size=batch_size)
    random_x = X_train[random_index]
    random_x = np.expand_dims(random_x, 3)
    random_y = train_labels[random_index]

    train_dict = {x_input: random_x, y_target: random_y}
```

```
sess.run(train_step, feed_dict=train_dict)
temp_train_loss, temp_train_preds = sess.run([loss,
prediction], feed_dict=train_dict)
temp_train_acc = get_accuracy(temp_train_preds, random_y)

eval_index = np.random.choice(len(X_test),
size=evaluation_size)
eval_x = X_test[eval_index]
eval_x = np.expand_dims(eval_x, 3)
eval_y = test_labels[eval_index]
test_dict = {eval_input: eval_x, eval_target: eval_y}
test_preds = sess.run(test_prediction, feed_dict=test_dict)
temp_test_acc = get_accuracy(test_preds, eval_y)
```

The results of the model are recorded in the following format and printed in the output:

```
# Record and print results
train_loss.append(temp_train_loss)
train_acc.append(temp_train_acc)
test_acc.append(temp_test_acc)
print('Epoch # {}. Train Loss: {:.2f}. Train Acc : {:.2f} .
temp_test_acc : {:.2f}'.format(i+1,temp_train_loss,
temp_train_acc,temp_test_acc))
```

Using a High-Level API for Building CNN Models

TFLearn, TensorLayer, tflayers, TF-Slim, tf.contrib.learn, Pretty Tensor, keras, and Sonnet are high-level TensorFlow APIs. If you use any of these high-level APIs, you can build CNN models in a few lines of code. So, you can explore any of these APIs for working smartly.

CHAPTER 8

CNN in Keras

This chapter will demonstrate how to use Keras to build CNN models. A CNN model can help you build an image classifier that can predict and classify the images. In general, you create some layers in the model architecture with initial values of weight and bias. Then you tune the weight and bias variables with the help of a training data set. You will learn how to code in Keras in this context. There is another approach that involves using pretrained models such as InceptionV3 and ResNet50 that can classify the images.

Let's define a CNN model and evaluate how well it performs. You will use a structure with a convolutional layer; then you will use max pooling and flatten out the network to fully connect the layers and make predictions.

Building an Image Classifier for MNIST Data in Keras

Here I will demonstrate the process of building a classifier for handwritten digits using the popular MNIST data set.

This task is a big challenge for playing with neural networks, but it can be managed on a single computer.

The MNIST database contains 60,000 training images and 10,000 testing images.

© Navin Kumar Manaswi 2018
N. K. Manaswi, *Deep Learning with Applications Using Python*,
https://doi.org/10.1007/978-1-4842-3516-4_8

Start by importing Numpy and setting a seed for the computer's pseudorandom number generator. This allows you to reproduce the results from your script.

```
import numpy as np
# random seed for reproducibility
np.random.seed(123)
```

Next, you import the sequential model type from Keras. This is simply a linear stack of neural network layers.

```
from keras.models import Sequential
from keras.layers import Dense
from keras.layers import Dropout
from keras.layers import Flatten
from keras.layers import Conv2D
from keras.layers import MaxPooling2d
#Now we will import some utilities
from keras.utils import np_utils
#Fixed dimension ordering issue
from keras import backend as K
K.set_image_dim_ordering('th')
#Load image data from MNIST
#Load pre-shuffled MNIST data into train and test sets
(X_train,y_train),(X_test, y_test)=mnist.load_data()

#Preprocess imput data for Keras
# Reshape input data.
# reshape to be [samples][channels][width][height]
X_train=X_train.reshape(X_train.shape[0],1,28,28)
X_test=X_test.reshape(X_test.shape[0],1,28,28)
```

```
# to convert our data type to float32 and normalize our database
X_train=X_train.astype('float32')
X_test=X_test.astype('float32')
print(X_train.shape)

# Z-scoring or Gaussian Normalization
X_train=X_train - np.mean(X_train) / X_train.std()
X_test=X_test - np.mean(X_test) / X_test.std()
#(60000, 1, 28, 28)

# convert 1-dim class arrays to 10 dim class metrices
#one hot encoding outputs
y_train=np_utils.to_categorical(y_train)
y_test-np_utils.to_categorical(y_test)
num_classes=y_test.shape[1]
print(num_classes)
#10

#Define a simple CNN model
print(X_train.shape)
#(60000,1,28,28)
```

Define the Network Structure

The network structure is as follows:

- Network has a convolutional input layer, with 32 feature maps with a size of 5×5. The activation function is rectified linear unit.

- The max pool layer has a size of 2×2.

- The dropout is set to 30 percent.

- You can flatten the layer.

- The network has a fully connected layer with 240 units, and the activation function is an exponential linear unit.

- Last layer of the netowrk is a fully connected output layer with ten units, and the activation function is softmax.

Then you compile the model by using binary cross entropy as the loss function and adagrad as the optimizer.

Define the Model Architecture

The architecture consists of a combination of the convolutional layer and max pooling layer and a dense layer at the end.

 # create a model

```
model=Sequential()
model.add(Conv2D(32, (5,5), input_shape=(1,28,28),
activation='relu'))
model.add(MaxPooling2D(pool_size=(2,2)))
model.add(Dropout(0.3))        # Dropout, one form of
regularization
model.add(Flatten())
model.add(Dense(240,activation='elu'))
model.add(Dense(num_classes, activation='softmax'))
print(model.output_shape)
(None, 10)
```

 # Compile the model

```
model.compile(loss='binary_crossentropy', optimizer='adagrad',
matrices=['accuracy'])
```

Then you fit the model by using the training data sets by taking a batch size of 200. The model takes first 200 instances/rows (from the 1st to the 200th) from the training data set and trains the network. Then the model takes second 200 instances (from the 201st to the 400th) for the training network again. In this way, you propagate all instances through the networks. The model requires less memory as you train networks with fewer instances each time. But the small batch size doesn't offer a good estimate of the gradient, and hence tuning the weight and bias can be challenge.

One epoch means one forward pass and one backward pass of *all* the training examples. It takes several iterations to complete one epoch.

Here, you have 60,000 training examples, and your batch size is 200, so it will take 300 iterations to complete 1 epoch.

```
# Fit the model
model.fit(X_train, y_train, validation_data=(X_test, y_test),
epochs=1, batch_size=200)
```

 # Evaluate model on test data
```
    # Final evaluation of the model
    scores =model.evaluate(X_test, y_test, verbose=0)
    print("CNN error: % .2f%%" % (100-scores[1]*100))
    # CNN Error: 17.98%

    # Save the model
    # save model
    model_json= model.to_join()
    with open("model_json", "w") as json_file:
    json_file.write(model_json)
    # serialize weights to HDFS
    model.save_weights("model.h5")
```

Building an Image Classifier with CIFAR-10 Data

This section explains how you can build a classifier that can classify the ten labels of the CIFAR-10 data set using the Keras CNN model.

Note   The CIFAR-10 data set consists of 60,000 32×32 color images in 10 classes, with 6,000 images per class. There are 50,000 training images and 10,000 test images.

```
###########Building CNN Model with CIFAR10 data##################
# plot cifar10 instances
    from keras.datasets import cifar10
    from matplotlib import pyplot
    from scipy.misc import toimage
    import numpy
    from keras.models import Sequential
    from keras.layers import Dense
    from keras.layers import Dropout
    from keras.layers import Flatten
    from keras.layers import Conv2D
    from keras.layers import MaxPooling2d
    #Now we will import some utilities
    from keras.utils import np_utils
    from keras.layers.normalization import BatchNormalization

    #Fixed dimension ordering issue
    from keras import backend as K
    K.set_image_dim_ordering('th')

    # fix random seed for reproducibility
    seed=12
```

```
numpy.random.seed(seed)
#Preprocess imput data for Keras
# Reshape input data.
# reshape to be [samples][channels][width][height]
X_train=X_train.reshape(X_train.shape[0],3,32,32).
astype('float32')
X_test=X_test.reshape(X_test.shape[0],3,32,32).
astype('float32')

# Z-scoring or Gaussian Normalization
X_train=X_train - np.mean(X_train) / X_train.std()
X_test=X_test - np.mean(X_test) / X_test.std()

# convert 1-dim class arrays to 10 dim class metrices
#one hot encoding outputs
y_train=np_utils.to_categorical(y_train)
y_test-np_utils.to_categorical(y_test)
num_classes=y_test.shape[1]
print(num_classes)
#10

#Define a simple CNN model
print(X_train.shape)
#(50000,3,32,32)
```

Define the Network Structure

The network structure is as follows:

- The convolutional input layer has 32 feature maps with a size of 5×5, and the activation function is a rectified linear unit.

- The max pool layer has a size of 2×2.

- The convolutional layer has 32 feature maps with a size of 5×5, and the activation function is a rectified linear unit.

- The network has batch normalization.

- The max pool layer has a size of 2×2.

- The dropout is set to 30 percent.

- You can flatten the layer.

- The fully connected layer has 240 units, and the activation function is an exponential linear unit.

- The fully connected output layer has ten units, and the activation function is softmax.

Then you fit the model by using the training data sets by taking a batch size of 200. You take the first 200 instances/rows (from the 1st to the 200th) from the training data set and train the network. Then you take the second 200 instances (from the 201st to the 400th) to train the network again. In this way, you propagate all instances through the networks. One epoch means one forward pass and one backward pass of *all* the training examples. It takes several iterations to complete one epoch.

Here, you have 50,000 training examples, and your batch size is 200, so it will take 250 iterations to complete 1 epoch.

Define the Model Architecture

A sequential model is created with a combination of convolutional and max pooling layers. Later a fully connected dense layer is attached.

```
# create a model
model=Sequential()
model.add(Conv2D(32, (5,5), input_shape=(3,32,32),
activation='relu'))
```

```
model.add(MaxPooling2D(pool_size=(2,2)))
model.add(Conv2D(32, (5,5), activation='relu',
padding='same'))
model.add(BatchNormalization())
model.add(MaxPooling2D(pool_size=(2,2)))
model.add(Dropout(0.3))          # Dropout, one form of
regularization
model.add(Flatten())
model.add(Dense(240,activation='elu'))
model.add(Dense(num_classes, activation='softmax'))
print(model.output_shape)
model.compile(loss='binary_crossentropy', optimizer='adagrad')
# fit model
model.fit(X_train, y_train, validation_data=(X_test,
y_test), epochs=1, batch_size=200)

# Final evaluation of the model
scores =model.evaluate(X_test, y_test, verbose=0)
print("CNN error: % .2f%%" % (100-scores[1]*100
```

Pretrained Models

In this section, I will show how you can use pretrained models such as VGG and inception to build up a classifier.

```
from keras import applications
from keras,models import Sequential, Model
from keras.applications.vgg16 import VGG16
from keras.applications.vgg16 import preprocess_input,
decode_predictions
from keras.models import Model
```

```
model = VGG16(weights='imagenet', include_top=True)
model.summary()

#predicting for any new image based on the pre-trained model
# Loading Image
img = image.load_img('('./Data/horse.jpg', target_size=(224, 224))
img = image.img_to_array(img)
img = np.expand_dims(img, axis=0)
img=preprocess_input(img)

# Predict the output
preds = model.predict(img)

# decode the predictions
pred_class = decode_predictions(preds, top=3)[0][0]
print("Predicted Class: %s" %pred_class[1])
print("Confidence("Confidance: %s"% pred_class[2])
#Predicted Class: hartebeest
#Confidence: 0.964784
ResNet50 and InceptionV3 models can be easily utilized for
prediction/classification of new images.
from keras.applications import ResNet50
model = ResNet50(weights='imagenet' , include_top=True)
model.summary()

# create the base pre-trained model
from keras.applications import InceptionV3
model = InceptionV3(weights='imagenet')
model.summary
```

Inception-V3 pre-trained model can detect/classify objects of 22,000 categories. It can detect/classify tray, torch, umbrella and others.

In many scenario, we need to build classifiers as per our requirement. For that, transfer learning is used where we use pre-trained model (used for feature extraction) and multiple neural.

CHAPTER 9

RNN and LSTM

This chapter will discuss the concepts of recurrent neural networks (RNNs) and their modified version, long short-term memory (LSTM). LSTM is mainly used for sequence prediction. You will learn about the varieties of sequence prediction and then learn how to do time-series forecasting with the help of the LSTM model.

The Concept of RNNs

A *recurrent neural network* is a type of artificial neural network that is best suited to recognizing patterns in sequences of data, such as text, video, speech, language, genomes, and time-series data. An RNN is an extremely powerful algorithm that can classify, cluster, and make predictions about data, particularly time series and text.

RNN can be seen as an MLP network with addition of loops to the architecture. In Figure 9-1, you can see that there is an input layer (with nodes such as x1, x2, and so on), a hidden layer (with nodes such as h1, h2, and so on), and an output layer (with nodes such as y1, y2, and so on). This is similar to the MLP architecture. The difference is that the nodes of the hidden layers are interconnected. In a vanilla (basic) RNN/LSTM, nodes are connected in one direction. This means that h2 depends on h1 (and x2), and h3 depends on h2 (and x3). The node in the hidden layer is decided by the previous node in the hidden layer.

© Navin Kumar Manaswi 2018
N. K. Manaswi, *Deep Learning with Applications Using Python*,
https://doi.org/10.1007/978-1-4842-3516-4_9

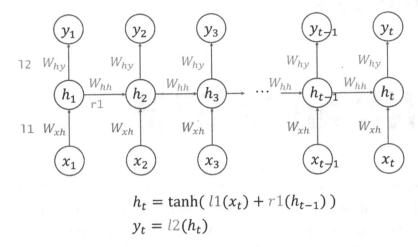

$$h_t = \tanh(\, l1(x_t) + r1(h_{t-1})\,)$$
$$y_t = l2(h_t)$$

Figure 9-1. *An RNN*

This kind of architecture ensures that the output at t=n depends on the inputs at t=n, t=n-1, ..., and t=1. In other words, the output depends on the sequence of data rather than a single piece of data (Figure 9-2).

(Input1) → Output1

(Input2, Input1)→ Output2

(Input3, Input2, Input1) → Output3

(Input4, Input3, Input2, Input1) → Output4

Figure 9-2. *The sequence*

Figure 9-3 shows how the nodes of the hidden layer are connected to the nodes of the input layer.

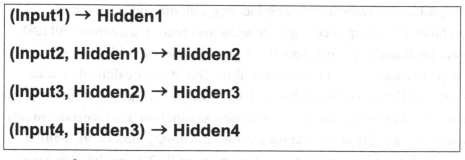

(Input1) → Hidden1

(Input2, Hidden1) → Hidden2

(Input3, Hidden2) → Hidden3

(Input4, Hidden3) → Hidden4

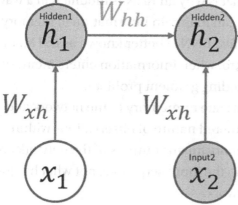

Figure 9-3. *The connections*

In an RNN, if the sequences are quite long, the gradients (which are essential for tuning the weight and bias) are computed during their training (backpropagation). They either vanish (multiplication of many small values less than 1) or explode (multiplication of many large values more than 1), causing the model to train very slowly.

The Concept of LSTM

Long short-term memory is a modified RNN architecture that tackles the problem of vanishing and exploding gradients and addresses the problem of training over long sequences and retaining memory. All RNNs have feedback loops in the recurrent layer. The feedback loops help keep information in "memory" over time. But, it can be difficult to train standard RNNs to solve problems that require learning long-term temporal dependencies. Since the gradient of the loss function decays exponentially with time (a phenomenon known as the *vanishing gradient problem*), it is difficult to train typical RNNs. That is why an RNN is modified in a way that it includes a memory cell that can maintain information in memory for long periods of time. The modified RNN is better known as LSTM. In LSTM, a set of gates is used to control when information enters memory, which solves the vanishing or exploding gradient problem.

The recurrent connections add state or memory to the network and allow it to learn and harness the ordered nature of observations within input sequences. The internal memory means outputs of the network are conditional on the recent context in the input sequence, not what has just been presented as input to the network.

Modes of LSTM

LSTM can have one of the following modes:

- One-to-one model

- One-to-many model

- Many-to-one model

- Many-to-many model

In addition to these modes, synced many-to-many models are also being used, especially for video classification.

118

Figure 9-4 shows a many-to-one LSTM. This implies that many inputs create one output in this model.

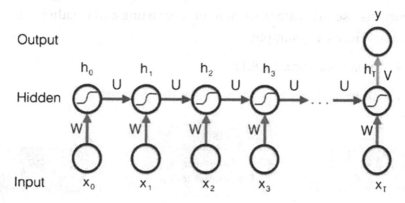

Figure 9-4. *Many-to-one LSTM*

Sequence Prediction

LSTM is best suited for sequence data. LSTM can predict, classify, and generate sequence data. A *sequence* means an order of observations, rather than a set of observations. An example of a sequence is a test series where the timestamps and values are in the order (chronologically) of the sequence. Another example is a video, which can be considered as a sequence of images or a sequence of audio clips.

Prediction based on the sequence of data is called the *sequence prediction*. Sequence prediction is said to have four types.

- Sequence numeric prediction

- Sequence classification

- Sequence generation

- Sequence-to-sequence prediction

119

Sequence Numeric Prediction

Sequence numeric prediction is predicting the next value for a given sequence. Its use cases are stock market forecasting and weather forecasting. Here's an example:

- *Input sequence*: 3,5,8,12

- *Output*: 17

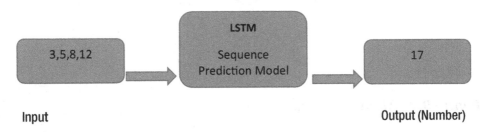

Input   Output (Number)

Sequence Classification

Sequence classification predicts the class label for a given sequence. Its use cases are fraud detection (which uses the transaction sequence as input to classify/predict whether an account has been hacked or not) and the classification of students based on performance (the sequence of exam marks over the last six months chronologically). Here's an example:

- *Input sequence*: 2,4,6,8

- *Output*: "Increasing"

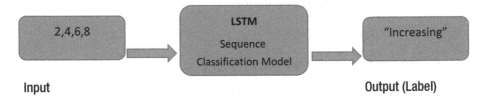

Input   Output (Label)

Sequence Generation

Sequence generation is when you generate a new output sequence that has the same properties as the input sequences in the input corpus. Its use cases are text generation (given 100 lines of a blog, generate the next line of the blog) and music generation (given the music examples, generate the new musical piece). Here's an example:

- *Input sequence*: [3, 5,8,12], [4,6,9,13]

- *Output*: [5,7,10,14]

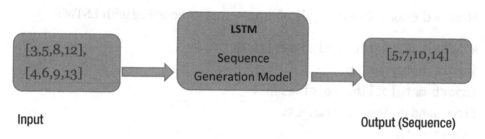

Input   Output (Sequence)

Sequence-to-Sequence Prediction

Sequence-to-sequence prediction is when you predict the next sequence for a given sequence. Its use cases are document summarization and multistep time-series forecasting (predicting a sequence of numbers). Here's an example:

- *Input sequence*: [3, 5,8,12,17]

- *Output*: [23,30,38]

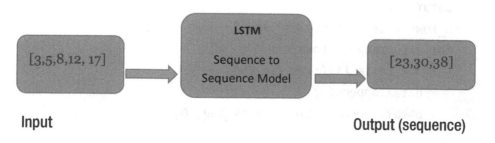

Input   Output (sequence)

As mentioned, LSTM is used for time-series forecasting in businesses.

Let's go through an LSTM model. Assume that a CSV file is given where the first column is a timestamp and the second column is a value. It can represent sensor (IoT) data.

Given the time-series data, you have to predict values for the future.

Time-Series Forecasting with the LSTM Model

Here is the complete example of time-series forecasting with LSTM:

```
# Simple LSTM for a time series data
import numpy as np
import matplotlib.pyplot as plt
from pandas import read_csv
import math
from keras.models import Sequential
from keras.layers import Dense
from keras.layers import LSTM
from sklearn.preprocessing import StandardScaler
from sklearn.metrics import mean_squared_error
import pylab

# convert an array of values into a timeseries data
def create_timeseries(series, ts_lag=1):
    dataX = []
    dataY = []
    n_rows = len(series)-ts_lag
    for i in range(n_rows-1):
        a = series[i:(i+ts_lag), 0]
        dataX.append(a)
        dataY.append(series[i + ts_lag, 0])
```

```
    X, Y = np.array(dataX), np.array(dataY)
    return X, Y
# fix random seed for reproducibility
np.random.seed(230)
# load dataset
dataframe = read_csv('sp500.csv', usecols=[0])
plt.plot(dataframe)
plt.show()
```

Figure 9-5 shows a plot of the data.

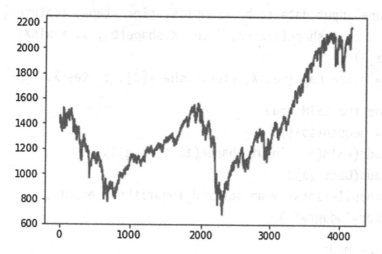

Figure 9-5. *Plot of the data*

Here's some more code:

```
# Changing datatype to float32 type
series = dataframe.values.astype('float32')

# Normalize the dataset
scaler = StandardScaler()
series = scaler.fit_transform(series)
```

```
# split the datasets into train and test sets
train_size = int(len(series) * 0.75)
test_size = len(series) - train_size
train, test = series[0:train_size,:], series[train_
size:len(series),:]

# reshape the train and test dataset into X=t and Y=t+1
ts_lag = 1
trainX, trainY = create_timeseries(train, ts_lag)
testX, testY = create_timeseries(test, ts_lag)

# reshape input data to be [samples, time steps, features]
trainX = np.reshape(trainX, (trainX.shape[0], 1, trainX.
shape[1]))
testX = np.reshape(testX, (testX.shape[0], 1, testX.shape[1]))

# Define the LSTM model
model = Sequential()
model.add(LSTM(10, input_shape=(1, ts_lag)))
model.add(Dense(1))
model.compile(loss='mean_squared_logarithmic_error',
optimizer='adagrad')

# fit the model
model.fit(trainX, trainY, epochs=500, batch_size=30)
# make predictions
trainPredict = model.predict(trainX)
testPredict = model.predict(testX)

# rescale predicted values
trainPredict = scaler.inverse_transform(trainPredict)
trainY = scaler.inverse_transform([trainY])
testPredict = scaler.inverse_transform(testPredict)
testY = scaler.inverse_transform([testY])
```

```
# calculate root mean squared error
trainScore = math.sqrt(mean_squared_error(trainY[0],
trainPredict[:,0]))
print('Train Score: %.2f RMSE' % (trainScore))
testScore = math.sqrt(mean_squared_error(testY[0],
testPredict[:,0]))
print('Test Score: %.2f RMSE' % (testScore))

# plot baseline and predictions
pylab.plot(trainPredictPlot)
pylab.plot(testPredictPlot)
pylab.show()
```

In Figure 9-6, you can see the plot of actual versus predicted time series. The part in orange is the training data, the part in blue is the test data, and the part in green is the predicted output.

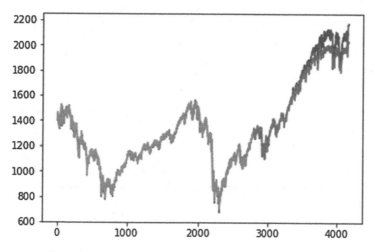

Figure 9-6. *Plot of actual versus predicted time series*

So far, we have learnt the concepts of RNN, LSTM and time series forecasting with LSTM model.

LSTM has been used in text classification. We use LSTM (vanilla LSTM or bi-directional LSTM) for building text classifiers. First, text corpus is converted into numbers by using word (semantic) embedding such as word2vec or glove. Then, sequence classification is done through LSTM. This approach offers much more accuracy than typical bag of words or tf-idf followed by ML classifiers such as SVM, Random Forest. In chapter 11, we can see how LSTM can be used for classifiers.

CHAPTER 10

Speech to Text and Vice Versa

In this chapter, you will learn about the importance of speech-to-text and text-to-speech conversion. You will also learn about the functions and components needed to do this type of conversion.

Specifically, I will cover the following:

- Why you would want to convert speech to text

- Speech as data

- Speech features that map speech to a matrix

- Spectrograms, which map speech to an image

- Building a classifier for speech recognition through mel-frequency cepstral coefficient (MFCC) features

- Building a classifier for speech recognition through spectrograms

- Open source approaches for speech recognition

- Popular cognitive service providers

- The future of speech of text

© Navin Kumar Manaswi 2018
N. K. Manaswi, *Deep Learning with Applications Using Python*,
https://doi.org/10.1007/978-1-4842-3516-4_10

Speech-to-Text Conversion

Speech-to-text conversion, in layman's terms, means that an app recognizes the words spoken by a person and converts the voice to written text. There are lots of reasons you would want to use Speech-to-Text conversion.

- Blind or physically challenged people can control different devices using only voice.

- You can keep records of meetings and other events by converting the spoken conversation to text transcripts.

- You can convert the audio in video and audio files to get subtitles of the words being spoken.

- You can translate words into another language by speaking into a device in one language and converting the text to speech in another language.

Speech as Data

The first step of making any automated speech recognition system is to get the *features*. In other words, you identify the components of the audio wave that are useful for recognizing the linguistic content and delete all the other useless features that are just background noises.

Each person's speech is filtered by the shape of their vocal tract and also by the tongue and teeth. What sound is coming out depends on this shape. To identify the phoneme being produced accurately, you need to determine this shape accurately. You could say that the shape of the vocal tract manifests itself to form an envelope of the short-time power spectrum. It's the job of MFCCs to represent this envelope accurately.

Speech can also be represented as data by converting it to a spectrogram (Figure 10-1).

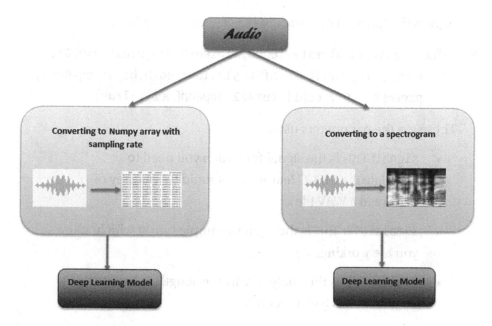

Figure 10-1. *Speech as data*

Speech Features: Mapping Speech to a Matrix

MFCCs are widely used in automated speech and speaker recognition. The mel scale relates the perceived frequency, or *pitch*, of a pure tone to its actual measured frequency.

You can convert an audio in frequency scale to the mel scale using the following formula:

$$M(f) = 1125 \ln(1 + f/700)$$

To convert it back to frequency, use the following formula:

$$M^{-1}(m) = 700(\exp(m/1125) - 1)$$

Here is the function to extract MFCC features in Python:

```
def mfcc(signal,samplerate=16000,winlen=0.025,winstep=0.01,
        numcep=13, nfilt=26,nfft=512,lowfreq=0,highfreq=None,
        preemph=0.97, ceplifter=22,appendEnergy=True)
```

These are the parameters used:

- signal: This is the signal for which you need to calculate the MFCC features. It should be an array of N*1 (read a WAV file).

- samplerate: This is the signal's sample rate at which you are working.

- winlen: This is the analysis window length in seconds. By default it is 0.025 second.

- winstep: This is the successive window step. By default it is 0.01 second.

- numcep: This is the number of ceptrum that the function should return. By default it is 13.

- nfilt: This is the number of filters in the filter bank. By default it is 26.

- nfft: This is the size of the fast Fourier transform (FFT). By default it is 512.

- lowfreq: This is the lowest band edge, in hertz. By default it is 0.

- highfreq: This is the highest band edge, in hertz. By default it is the sample rate divided by 2.

- preemph: This applies a preemphasis filter with preemph as the coefficient. 0 means no filter. By default it is 0.97.

- ceplifter: This applies a lifter to the final cepstral coefficients. 0 means no lifter. By default it is 22.

- appendEnergy: The zeroth cepstral coefficient is replaced with the log of the total frame energy, if it is set to true.

This function returns a Numpy array containing features. Each row contains one feature vector.

Spectrograms: Mapping Speech to an Image

A *spectrogram* is photographic or electronic representation of a spectrum. The idea is to convert an audio file into images and pass the images into deep learning models such as a CNN and LSTM for analysis and classification.

The spectrogram is computed as a sequence of FFTs of windowed data segments. A common format is a graph with two geometric dimensions; one axis represents time, and another axis represents frequency. A third dimension uses the color or size of point to indicate the amplitude of a particular frequency at a particular time. Spectrograms are usually created in one of two ways. They can be approximated as a filter bank that results from a series of band-pass filters. Or, in Python, there is a direct function that maps audio to a spectrogram.

Building a Classifier for Speech Recognition Through MFCC Features

To build a classifier for speech recognition, you need to have the python_speech_features Python package installed.

You can use the command `pip install python_speech_features` to install this package.

The `mfcc` function creates a feature matrix for an audio file. To build a classifier that recognizes the voices of different people, you need to collect speech data of them in WAV format. Then you convert all the audio files into a matrix using the `mfcc` function. The code to extract the features from the WAV file is shown here:

```
from python_speech_features import mfcc
from python_speech_features import delta
from python_speech_features import logfbank
import scipy.io.wavfile as wav

(samplerate,signal) = wav.read("audio.wav")
mfccfeatures = mfcc(signal,samplerate)
dmfccfeature = delta(mfccfeatures, 2)
fbankfeature = logfbank(signal,samplerate)

print(fbankfeature)
```

If you run the previous code, you will get output in the following form:

```
[[ 7.66608682  7.04137131  7.30715423 ...,  9.43362359  9.11932984
   9.93454603]
 [ 4.9474559   4.97057377  6.90352236 ...,  8.6771281   8.86454547
   9.7975147 ]
 [ 7.4795622   6.63821063  5.98854983 ...,  8.78622734  8.805521
   9.83712966]
 ...,
 [ 7.8886269   6.57456605  6.47895433 ...,  8.62870034  8.79965464
   9.67997298]
```

```
[ 5.73028657   4.87985847   6.64977329 ...,   8.64089442   8.62887745
  9.90470194]
[ 8.8449656    6.67098127   7.09752316 ...,   8.84914694   8.97807983
  9.45123015]]
```

Here, each row represents one feature vector.

Collect as many voice recordings of a person as you can and append the feature matrix of each audio file in this matrix.

This will act as your training data set.

Repeat the same steps with all the other classes.

Once the data set is prepared, you can fit this data into any deep learning model (that is used for classification) to classify the voices of different people.

Note   To see the full code of a classifier using MFCC features, you can visit www.navinmanaswi.com/SpeechRecognizer.

Building a Classifier for Speech Recognition Through a Spectrogram

Using the spectrogram approach converts all the audio files to images (Figure 10-2), so all you have to do is convert all the sound files in the training data into images and feed those images to a deep learning model just like you do in a CNN.

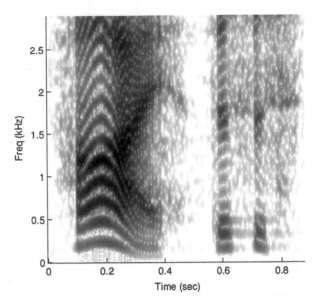

Figure 10-2. *Spectogram of speech sample*

Here is the Python code to convert an audio file to a spectrogram:

```python
import matplotlib.pyplot as plt
from scipy import signal
from scipy.io import wavfile

sample_rate, samples = wavfile.read('monoAudioFile.wav')
frequencies, times, spectogram = signal.spectrogram(samples, sample_rate)

plt.imshow(spectogram)
plt.ylabel('Freq(kHz)')
plt.xlabel('Time (sec)')
plt.show()
```

Open Source Approaches

There are open source packages available for Python that perform speech-to-text and text-to-speech conversion.

The following are some open source speech-to-text conversion APIs:

- PocketSphinx
- Google Speech
- Google Cloud Speech
- Wit.ai
- Houndify
- IBM Speech to Text API
- Microsoft Bing Speech

Having used all of these, I can say that they work quite well; the American accent is especially clear.

If you are interested in evaluating the accuracy of the conversion, you need one metric: the word error rate (WER).

In the next section, I will discuss each API mentioned previously.

Examples Using Each API

Let's go through each API.

Using PocketSphinx

PocketSphinx is an open source API used for speech-to-text conversions. It is a lightweight speech recognition engine, specifically tuned for handheld and mobile devices, though it works equally well on the desktop. Simply use the command `pip install PocketSphinx` to install the package.

```
import speech_recognition as sr
from os import path
AUDIO_FILE = "MyAudioFile.wav"

r = sr.Recognizer()
with sr.AudioFile(AUDIO_FILE) as source:
 audio = r.record(source)

try:
    print("Sphinx thinks you said " + r.recognize_sphinx(audio))
except sr.UnknownValueError:
    print("Sphinx could not understand audio")
except sr.RequestError as e:
  print("Sphinx error; {0}".format(e))
```

==

Using the Google Speech API

Google provides its own Speech API that can be implemented in Python code and can be used to create different applications.

```
# recognize speech using Google Speech Recognition
try:
    print("Google Speech Recognition thinks you said " +
    r.recognize_google(audio))
except sr.UnknownValueError:
    print("Google Speech Recognition could not understand audio")
except sr.RequestError as e:
    print("Could not request results from Google Speech
    Recognition service;{0}".format(e))
```

Using the Google Cloud Speech API

You can also use the Google Cloud Speech API for the conversion. Create an account on the Google Cloud and copy the credentials.

```
GOOGLE_CLOUD_SPEECH_CREDENTIALS = r"INSERT THE CONTENTS OF THE
GOOGLE CLOUD SPEECH JSON CREDENTIALS FILE HERE" try:
    print("Google Cloud Speech thinks you said " +
    r.recognize_google_cloud(audio, credentials_json=GOOGLE_
    CLOUD_SPEECH_CREDENTIALS))
except sr.UnknownValueError:
    print("Google Cloud Speech could not understand audio")
except sr.RequestError as e:
    print("Could not request results from Google Cloud Speech
    service; {0}".format(e))
```

Using the Wit.ai API

The Wit.ai API enables you to make a speech-to-text converter. You need to create an account and then create a project. Copy your Wit.ai key and start coding.

```
#recognize speech using Wit.ai
WIT_AI_KEY = "INSERT WIT.AI API KEY HERE" # Wit.ai keys are
32-character uppercase alphanumeric strings
try:
    print("Wit.ai thinks you said " + r.recognize_wit(audio,
    key=WIT_AI_KEY))
except sr.UnknownValueError:
    print("Wit.ai could not understand audio")
except sr.RequestError as e:
    print("Could not request results from Wit.ai service; {0}".
    format(e))
```

Using the Houndify API

Similar to the previous APIs, you need to create an account at Houndify and get your client ID and key. This allows you to build an app that responds to sound.

```
# recognize speech using Houndify
HOUNDIFY_CLIENT_ID = "INSERT HOUNDIFY CLIENT ID HERE"
# Houndify client IDs are Base64-encoded strings
HOUNDIFY_CLIENT_KEY = "INSERT HOUNDIFY CLIENT KEY HERE"
# Houndify client keys are Base64-encoded strings
try:
    print("Houndify thinks you said " + r.recognize_
    houndify(audio, client_id=HOUNDIFY_CLIENT_ID, client_
    key=HOUNDIFY_CLIENT_KEY))
except sr.UnknownValueError:
    print("Houndify could not understand audio")
except sr.RequestError as e:
    print("Could not request results from Houndify service;
    {0}".format(e))
```

Using the IBM Speech to Text API

The IBM Speech to Text API enables you to add IBM's speech recognition capabilities to your applications. Log in to the IBM cloud and start your project to get an IBM username and password.

```
# IBM Speech to Text
# recognize speech using IBM Speech to Text
IBM_USERNAME = "INSERT IBM SPEECH TO TEXT USERNAME HERE" # IBM
Speech to Text usernames are strings of the form XXXXXXXX-XXXX-
XXXX-XXXX-XXXXXXXXXXXX
```

```
IBM_PASSWORD = "INSERT IBM SPEECH TO TEXT PASSWORD HERE" # IBM
Speech to Text passwords are mixed-case alphanumeric strings
try:
    print("IBM Speech to Text thinks you said " + r.recognize_
    ibm(audio, username=IBM_USERNAME, password=IBM_PASSWORD))
except sr.UnknownValueError:
    print("IBM Speech to Text could not understand audio")
except sr.RequestError as e:
    print("Could not request results from IBM Speech to Text
    service; {0}".format(e))
```

Using the Bing Voice Recognition API

This API recognizes audio coming from a microphone in real time. Create an account on Bing.com and get a Bing Voice Recognition API key.

```
# recognize speech using Microsoft Bing Voice Recognition
BING_KEY = "INSERT BING API KEY HERE" # Microsoft Bing Voice
Recognition API key is 32-character lowercase hexadecimal
strings
try:
    print("Microsoft Bing Voice Recognition thinks you said " +
    r.recognize_bing(audio, key=BING_KEY))
except sr.UnknownValueError:
    print("Microsoft Bing Voice Recognition could not
    understand audio")
except sr.RequestError as e:
    print("Could not request results from Microsoft Bing Voice
    Recognition service; {0}".format(e))
```

Once you have converted the speech into text, you cannot expect 100 percent accuracy. To measure the accuracy, you can use the WER.

Text-to-Speech Conversion

This section of the chapter focuses on converting written text to an audio file.

Using pyttsx

Using a Python package called pyttsx, you can convert text to audio.

Do a pip install pyttsx. If you are using python 3.6 then do pip3 install pyttsx3.

```
import pyttsx
engine = pyttsx.init()
engine.say("Your Message")
engine.runAndWait()
```

Using SAPI

You can also use SAPI to do text-to-speech conversion in Python.

```
from win32com.client import constants, Dispatch
Msg = "Hi this is a test"
speaker = Dispatch("SAPI.SpVoice")  #Create SAPI SpVoice Object
speaker.Speak(Msg)                  #Process TTS
del speaker
```

Using SpeechLib

You can take the input from a text file and convert it to audio using SpeechLib, as shown here:

```
from comtypes.client import CreateObject
engine = CreateObject("SAPI.SpVoice")
```

```
stream = CreateObject("SAPI.SpFileStream")
from comtypes.gen import SpeechLib
infile = "SHIVA.txt"
outfile = "SHIVA-audio.wav"
stream.Open(outfile, SpeechLib.SSFMCreateForWrite)
engine.AudioOutputStream = stream
f = open(infile, 'r')
theText = f.read()
f.close()
engine.speak(theText)
stream.Close()
```

Many times, you have to edit the audio so that you can remove a voice from the audio file. The next section shows you how.

Audio Cutting Code

Make a CSV file of audio that contains the comma-separated values of the details of the audio and perform the following using Python:

```
import wave
import sys
import os
import csv
origAudio = wave.open('Howard.wav', 'r') #change path
frameRate = origAudio.getframerate()
nChannels = origAudio.getnchannels()
sampWidth = origAudio.getsampwidth()
nFrames   = origAudio.getnframes()

filename =  'result1.csv' #change path
```

```
exampleFile = open(filename)
exampleReader = csv.reader(exampleFile)
exampleData = list(exampleReader)

count = 0

for data in exampleData:
 #for selections in data:
    print('Selections ', data[4], data[5])
    count += 1
    if data[4] == 'startTime' and data[5] == 'endTime':
        print('Start time')
    else:
        start = float(data[4])
        end = float(data[5])
        origAudio.setpos(start*frameRate)
        chunkData = origAudio.readframes(int((end-
        start)*frameRate))
        outputFilePath = 'C:/Users/Navin/outputFile{0}.wav'.
        format(count) # change path
        chunkAudio = wave.open(outputFilePath, 'w')
        chunkAudio.setnchannels(nChannels)
        chunkAudio.setsampwidth(sampWidth)
        chunkAudio.setframerate(frameRate)
        chunkAudio.writeframes(chunkData)
        chunkAudio.close()
```

Cognitive Service Providers

Let's look at some cognitive service providers that help with speech
processing.

Microsoft Azure

Microsoft Azure provides the following:

- *Custom Speech Service*: This overcomes speech recognition barriers such as speaking style, vocabulary, and background noise.

- *Translator Speech API*: This enables real-time speech translation.

- *Speaker Identification API*: This can identify the speakers based on a speech sample of each speaker in the given audio data.

- *Bing Speech API*: This converts audio to text, understands intent, and converts text back to speech for natural responsiveness.

Amazon Cognitive Services

Amazon Cognitive Services provides Amazon Polly, a service that turns text into speech. Amazon Polly lets you create applications that talk, enabling you to build entirely new categories of speech-enabled products.

- 47 voices and 24 languages can be used, and an Indian English option is provided.

- Tones such as whispering, anger, and so on, can be added to particular parts of the speech using Amazon effects.

- You can instruct the system how to pronounce a particular phrase or word in a different way. For example, "W3C" is pronounced as World Wide Web Consortium, but you can change that to pronounce just the acronym. You can also provide the input text in SSML format.

IBM Watson Services

There are two services from IBM Watson.

- *Speech to text*: U.S. English, Spanish, and Japanese

- *Text to speech*: U.S. English, U.K. English, Spanish, French, Italian, and German

The Future of Speech Analytics

Speech recognition technology has been making a great progress. Every year, it is about 10 to 15 percent more accurate than the previous year. In the future, it will provide the most interactive interface for computers yet.

There are many applications that you will soon be witnessing in the marketplace, including interactive books, robotic control, and self-driving car interfaces. Speech data offers some exciting new possibilities because it is the future of the industry. Speech intelligence enables people to message, take or give orders, raise complaints and to do any work where they used to type manually. It offers a great customer experience and perhaps that is why all customer-facing departments and businesses tend to use speech applications very heavily. I can see a great future for speech application developers.

CHAPTER 11

Developing Chatbots

Artificial intelligence systems that act as interfaces for human and machine interactions through text or voice are called *chatbots*.

The interactions with chatbots may be either straightforward or complex. An example of a straightforward interaction could be asking about the latest news report. The interactions can become more complex when they are about troubleshooting a problem with, say, your Android phone. The term *chatbots* has gained immense popularity in the past year and has grown into the most preferred platform for user interaction and engagement. A *bot*, an advanced form of a chatbot, helps automate "user-performed" tasks.

This chapter on chatbots will serve as an all-encompassing guide to the what, how, where, when, and why of chatbots!

Specifically, I will cover the following:

- Why you would want to use chatbots

- The designs and functions of chatbots

- The steps for building a chatbot

- Chatbot development using APIs

- The best practices of chatbots

© Navin Kumar Manaswi 2018
N. K. Manaswi, *Deep Learning with Applications Using Python*,
https://doi.org/10.1007/978-1-4842-3516-4_11

Why Chatbots?

It is important for a chatbot to understand what information a user is seeking, called the *intent*. Suppose a user wants to know the nearest vegetarian restaurant; the user can ask that question in many possible ways. A chatbot (specifically the intent classifier inside the chatbot) must be able to understand the intent because the user wants to get the right answer. In fact, to give the right answer, the chatbot must be able to understand the context, intent, entities, and sentiment. The chatbot has to even take account of whatever is discussed in the session. For example, the user might ask the question "What is the price of chicken biryani there?" Though the user has asked for a price, the chat engine can misunderstand and assume the user is looking for a restaurant. So, in response, the chatbot may provide the name of the restaurant.

Designs and Functions of Chatbots

A chatbot stimulates intelligent conversations with humans through the application of AI.

The interface through which conversation takes place is facilitated via either spoken or written text. Facebook Messenger, Slack, and Telegram make use of chatbot messaging platforms. They serve many purposes, including ordering products online, investing and managing finances, and so on. An important aspect of chatbots is that they make contextual conversation a possibility. The chatbots converse with users in a way similar to how human beings converse in their daily lives. Though it is possible for chatbots to converse contextually, they still have a long way to go in terms of communicating contextually with everything and anything. But chat interfaces are making use of language to connect the machine to the man, helping people get things done in a convenient manner by providing information in a contextual manner.

Moreover, chatbots are redefining the way businesses are being conducted. From reaching out to the consumers to welcoming them to the ecosystem of the business to providing information to the consumers about various products and their features, chatbots are helping with it all. They are emerging as the most convenient way of dealing with consumers in a timely and satisfactory manner.

Steps for Building a Chatbot

A chatbot is built to communicate with users and give them the feeling that they are communicating with a human and not a bot. But when users are giving input, it is common that they will not give input in the proper way. In other words, they may enter unnecessary punctuation marks, or there may be different ways of asking the same question.

For example, for "Restaurants near me?" a user could input "Restaurants beside me?" or "Find a nearby restaurant."

Therefore, you need to preprocess the data so that the chatbot engine can easily understand it. Figure 11-1 shows the process, which is detailed in the following sections.

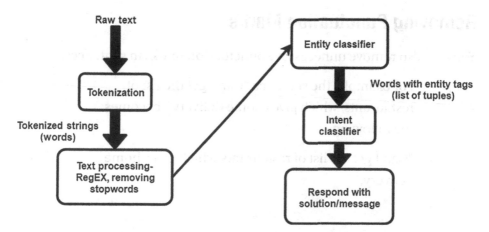

Figure 11-1. *A flowchart to show how a chatbot engine processes an input string and gives a valid reply.*

Preprocessing Text and Messages

Preprocessing text and messages includes several steps, covered next.

Tokenization

Chopping up sentences into single words (called *tokens*) is called *tokenization*. In Python, generally a string is tokenized and stored in a list.

For example, the sentence "Artificial intelligence is all about applying mathematics" becomes the following:

["Artificial", "intelligence", "is", "all", "about", "applying", "mathematics"]

Here is the example code:

```
from nltk.tokenize import TreebankWordTokenizer
l = "Artificial intelligence is all about applying mathematics"
token = TreebankWordTokenizer().tokenize(l)
print(token)
```

Removing Punctuation Marks

You can also remove unnecessary punctuation marks in sentences.

For example, the sentence "Can I get the list of restaurants, which gives home delivery." becomes the following:

"Can I get the list of restaurants which gives home delivery."

Here is the example code:

```
from nltk.tokenize import TreebankWordTokenizer
from nltk.corpus import stopwords
l = "Artificial intelligence is all about applying
mathematics!"
token = TreebankWordTokenizer().tokenize(l)
output = []
output = [k for k in token if k.isalpha()]
print(output)
```

Removing Stop Words

Stop words are the words present in a sentence that don't make much difference if removed. Though the format of the sentence changes, this helps a lot in natural language understanding (NLU).

For example, the sentence "Artificial intelligence can change the lifestyle of the people." becomes the following after removing stop words:

"Artificial intelligence change lifestyle people."

Here is the example code:

```
from nltk.tokenize import TreebankWordTokenizer
from nltk.corpus import stopwords
l = "Artificial intelligence is all about applying mathematics"
token = TreebankWordTokenizer().tokenize(l)
stop_words = set(stopwords.words('english'))
output= []
for k in token:
    if k not in stop_words:
        output.append(k)
print(output)
```

Which words are considered as the stop words can vary. There are some predefined sets of stop words provided by Natural Language Toolkit (NLTK), Google, and more.

Named Entity Recognition

Named entity recognition (NER), also known as *entity identification*, is the task of classifying entities in text into predefined classes such as the name of a country, the name of a person, and so on. You can also define your own classes.

For example, applying NER to the sentence "Today's India vs Australia cricket match was fantastic." gives you the following output:

[Today's]Time [India] Country vs [Australia] Country [cricket] Game match was fantastic.

To run the code for NER, you need to download and import the necessary packages, as mentioned in the following code.

Using Stanford NER

To run the code, download english.all.3class.distsim.crf.ser.gz and stanford-ner.jar files.

```
from nltk.tag import StanfordNERTagger
from nltk.tokenize import word_tokenize

StanfordNERTagger("stanford-ner/classifiers/english.all.3class.
distsim.crf.ser.gz",
"stanford-ner/stanford-ner.jar")

text = "Ron was the founder of Ron Institute at New york"
text = word_tokenize(text)
ner_tags = ner_tagger.tag(text)

print(ner_tags)
```

Using MITIE NER (Pretrained)

Download the ner_model.dat file of MITIE to run the code.

```
from mitie.mitie import *
from nltk.tokenize import word_tokenize

print("loading NER model...")
ner = named_entity_extractor("mitie/MITIE-models/english/
ner_model.dat".encode("utf8"))

text = "Ron was the founder of Ron Institute at New york".
encode("utf-8")
text = word_tokenize(text)

ner_tags = ner.extract_entities(text)
print("\nEntities found:", ner_tags)

for e in ner_tags:
        range = e[0]
        tag = e[1]
        entity_text = " ".join(text[i].decode() for i in range)
        print( str(tag) + " : " + entity_text)
```

Using MITIE NER (Self-Trained)

Download the total_word_feature_extractor.dat file of MITIE
(https://github.com/mit-nlp/MITIE) to run the code.

```
from mitie.mitie import *

sample = ner_training_instance([b"Ron", b"was", b"the", b"founder",
b"of", b"Ron", b"Institute", b"at", b"New", b"York", b"."])

sample.add_entity(range(0, 1), "person".encode("utf-8"))
sample.add_entity(range(5, 7), "organization".encode("utf-8"))
sample.add_entity(range(8, 10), "Location".encode("utf-8"))
```

```
trainer = ner_trainer("mitie/MITIE-models/english/total_word_
feature_extractor.dat".encode("utf-8"))

trainer.add(sample)

ner = trainer.train()

tokens = [b"John", b"was", b"the", b"founder", b"of", b"John",
b"University", b"."]
entities = ner.extract_entities(tokens)
print ("\nEntities found:", entities)
for e in entities:
        range = e[0]
        tag = e[1]
        entity_text = " ".join(str(tokens[i]) for i in range)
        print ("      " + str(tag) + ": " + entity_text)
```

Intent Classification

Intent classification is the step in NLU where you try to understand what the user wants. Here are two examples of input to a chatbot to find places nearby:

- *"I need to buy groceries."*: The intent is to look for a grocery store nearby.

- *"I want to have vegetarian food."*: The intent is to look for restaurants nearby, ideally vegetarian ones.

Basically, you need to understand what the user is looking for and accordingly classify the request into a certain category of intent (Figure 11-2).

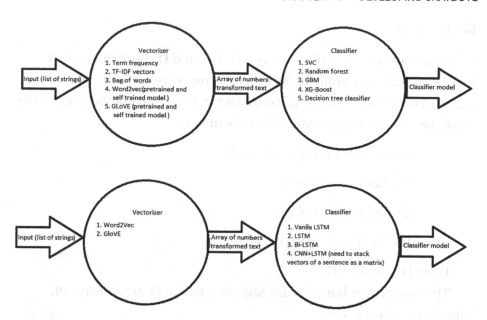

Figure 11-2. *General flow of intent classification, from sentences to vectors to a model*

To do this, you need to train a model to classify requests into intents using an algorithm, going from sentences to vectors to a model.

Word Embedding

Word embedding is the technique of converting text to numbers. It is difficult to apply any algorithm in text. Hence, you have to convert it to numbers.

The following are different types of word embedding techniques.

Count Vector

Suppose you have three documents (D1, D2, and D3) and there are N unique words in the group of documents. You create a (D×N) matrix, called C, which is known as the *count vector*. Each entry of the matrix is the frequency of the unique word in that document.

Let's see this using an example.

D1: Pooja is very lazy.

D2: But she is intelligent.

D3: She hardly comes to class.

Here, D=3 and N=12.

The unique words are *hardly, lazy, But, to, Pooja, she, intelligent, comes, very, class,* and *is*.

Hence, the count vector, C, will be the following:

	Hardly	laziest	But	to	Pooja	she	intelligent	comes	very	class	is
D1	0	1	0	0	1	0	0	0	1	0	1
D2	0	0	1	0	0	1	1	0	0	0	1
D3	1	0	0	1	0	1	0	1	0	1	0

Term Frequency-Inverse Document Frequency (TF-IDF)

For this technique, you give each word in the sentence a number depending upon how many times that word occurs in that sentence and also depending upon the document. Words occurring many times in a sentence and not so many times in a document will have high values.

For example, consider a set of sentences:

- "I am a boy."

- "I am a girl."

- "Where do you live?"

TF-IDF transforms the feature set for the previous sentences, as shown here:

	Am	Boy	Girl	Where	do	you	live
1.	0.60	0.80	0	0	0	0	0
2.	0.60	0	0.80	0	0	0	0
3.	0	0	0	0.5	0.5	0.5	0.5

You can import the TFIDF package and use it to create this table.

Now let's see some sample code. You can use a support vector classifier on the TF-IDF transformed features of the request string.

```
#import required packages
import pandas as pd
from random import sample
from sklearn.preprocessing import LabelEncoder
from sklearn.feature_extraction.text import TfidfVectorizer
from sklearn.svm import SVC
from sklearn.model_selection import train_test_split
from sklearn.metrics import f1_score, accuracy_score
# read csv file
data = pd.read_csv("intent1.csv")
print(data.sample(6))
```

Before continuing with the code, here's an example of the data set:

Description (Message)	intent_label (Target)
Good Non-Veg restaurant near me	0
I am looking for a hospital	1
Good hospital for Heart operation	1
International school for kids	2
Non-Veg restaurant around me	0
School for small Kids	2

In this example, these are the values to use:

- 0 means looking for a restaurant.

- 1 means looking for a hospital.

- 2 means looking for a school.

Now let's work on the data set.

```
# split dataset into train and test.
X_train, X_test, Y_train, Y_test = train_test_split(data
["Description"], data["intent_label"], test_size=3)
print(X_train.shape, X_test.shape, Y_train.shape, Y_test.shape)

# vectorize the input using tfidf values.
tfidf = TfidfVectorizer()
tfidf = tfidf.fit(X_train)
X_train = tfidf.transform(X_train)
X_test = tfidf.transform(X_test)

# label encoding for different categories of intents
le = LabelEncoder().fit(Y_train)
Y_train = le.transform(Y_train)
Y_test = le.transform(Y_test)
```

```
# other models like GBM, Random Forest may also be used
model = SVC()
model = model.fit(X_train, Y_train)
p = model.predict(X_test)
# calculate the f1_score. average="micro" since we want to
calculate score for multiclass.
# Each instance(rather than class(search for macro average))
contribute equally towards the scoring.
print("f1_score:", f1_score( Y_test, p, average="micro"))
print("accuracy_score:",accuracy_score(Y_test, p))
```

Word2Vec

There are different methods of getting word vectors for a sentence, but the main theory behind all the techniques is to give similar words a similar vector representation. So, words like *man* and *boy* and *girl* will have similar vectors. The length of each vector can be set. Examples of Word2vec techniques include GloVe and CBOW (n-gram with or without skip grams).

You can use Word2vec by training it for your own data set (if you have enough data for the problem), or you can use pretrained data. Word2vec is available on the Internet. Pretrained models have been trained on huge documents such as Wikipedia data, tweets, and so on, and they're almost always good for the problem.

An example of some techniques that you can use to train your intent classifier is to use a 1D-CNN on word vectors of the words in a sentence, appended in a list for each sentence.

```
# import required packages
from gensim.models import Word2Vec
import pandas as pd
import numpy as np
from keras.preprocessing.text import Tokenizer
```

```
from keras.preprocessing.sequence import pad_sequences
from keras.utils.np_utils import to_categorical
from keras.layers import Dense, Input, Flatten
from keras.layers import Conv1D, MaxPooling1D, Embedding, Dropout
from keras.models import Model

from sklearn.preprocessing import LabelEncoder
from sklearn.model_selection import train_test_split
from sklearn.metrics import f1_score, accuracy_score
# read data
data = pd.read_csv("intent1.csv")

# split data into test and train
X_train, X_test, Y_train, Y_test = train_test_split(data
["Description"], data["intent_label"], test_size=6)

# label encoding for different categories of intents
le = LabelEncoder().fit(Y_train)
Y_train = le.transform(Y_train)
Y_test = le.transform(Y_test)

# get word_vectors for words in training set
X_train = [sent for sent in X_train]
X_test = [sent for sent in X_test]
# by default genism.Word2Vec uses CBOW, to train word vecs.
We can also use skipgram with it
# by setting the "sg" attribute to number of skips we want.
# CBOW and Skip gram for the sentence "Hi Ron how was your
day?" becomes:
# Continuos bag of words: 3-grams {"Hi Ron how", "Ron how was",
"how was your" ...}
# Skip-gram 1-skip 3-grams: {"Hi Ron how", "Hi Ron was", "Hi
how was", "Ron how
# your", ...}
```

```
# See how: "Hi Ron was" skips over "how".
# Skip-gram 2-skip 3-grams: {"Hi Ron how", "Hi Ron was", "Hi
Ron your", "Hi was
# your", ...}
# See how: "Hi Ron your" skips over "how was".
# Those are the general meaning of CBOW and skip gram.
word_vecs = Word2Vec(X_train)
print("Word vectors trained")

# prune each sentence to maximum of 20 words.
max_sent_len = 20

# tokenize input strings
tokenizer = Tokenizer()
tokenizer.fit_on_texts(X_train)
sequences = tokenizer.texts_to_sequences(X_train)
sequences_test = tokenizer.texts_to_sequences(X_test)
word_index = tokenizer.word_index
vocab_size = len(word_index)

# sentences with less than 20 words, will be padded with zeroes
to make it of length 20
# sentences with more than 20 words, will be pruned to 20.
x = pad_sequences(sequences, maxlen=max_sent_len)
X_test = pad_sequences(sequences_test, maxlen=max_sent_len)

# 100 is the size of wordvec.
embedding_matrix = np.zeros((vocab_size + 1, 100))

# make matrix of each word with its word_vectors for the CNN model.
# so each row of a matrix will represent one word. There will
be a row for each word in
```

```
# the training set
for word, i in word_index.items():
        try:
            embedding_vector = word_vecs[word]
        except:
            embedding_vector = None
            if embedding_vector is not None:
                embedding_matrix[i] = embedding_vector
print("Embeddings done")
vocab_size = len(embedding_matrix)

# CNN model requires multiclass labels to be converted into one
hot ecoding.
# i.e. each column represents a label, and will be marked one
for corresponding label.
y = to_categorical(np.asarray(Y_train))

embedding_layer = Embedding(vocab_size,
                            100,
                            weights=[embedding_matrix],
                            input_length=max_sent_len,
                            trainable=True)
sequence_input = Input(shape=(max_sent_len,), dtype='int32')

# stack each word of a sentence in a matrix. So each matrix
represents a sentence.
# Each row in a matrix is a word(Word Vector) of a sentence.
embedded_sequences = embedding_layer(sequence_input)

# build the Convolutional model.
l_cov1 = Conv1D(128, 4, activation='relu')(embedded_sequences)
l_pool1 = MaxPooling1D(4)(l_cov1)
l_flat = Flatten()(l_pool1)
```

```
hidden = Dense(100, activation='relu')(l_flat)
preds = Dense(len(y[0]), activation='softmax')(hidden)
model = Model(sequence_input, preds)
model.compile(loss='binary_crossentropy',optimizer='Adam')

print("model fitting - simplified convolutional neural
network")
model.summary()

# train the model
model.fit(x, y, epochs=10, batch_size=128)

#get scores and predictions.
p = model.predict(X_test)
p = [np.argmax(i) for i in p]
score_cnn = f1_score(Y_test, p, average="micro")
print("accuracy_score:",accuracy_score(Y_test, p))
print("f1_score:", score_cnn)
```

The model fitting is a simplified convolutional neural network, as shown here:

Layer (Type)	Output Shape	Param #
input_20 (InputLayer)	(None, 20)	0
embedding_20 (Embedding)	(None, 20, 100)	2800
conv1d_19 (Conv1D)	(None, 17, 128)	51328
max_pooling1d_19 (MaxPooling)	(None, 4, 128)	0
flatten_19 (Flatten)	(None, 512)	0
dense_35 (Dense)	(None, 100)	51300
dense_36 (Dense)	(None, 3)	303

Here are the numbers of parameters:

- *Total parameters*: 105,731

- *Trainable parameters*: 105,731

- *Nontrainable parameters*: 0

Here are some important functions of Word2vec using the Gensim package:

- This is how you import Gensim and load the pretrained model:

```
import genism
#loading the pre-trained model
model = gensim.models.KeyedVectors.
load_word2vec_format('GoogleNews-vectors-
negative300.bin', binary=True)
```

- This is the pretrained model from Google for the English language, and it is of 300 dimensions.

- This is how to find the word vector of a word from a pretrained model:

```
# getting word vectors of a word
lion = model['lion']
print(len(lion))
```

- This is how to find the similarity index between two words:

```
#Calculating similarity index
print(model.similarity('King', 'Queen'))
```

- This is how to find an odd one out from the set of words:

```
#Choose odd one out
print(model.doesnt_match("Mango Grape Tiger
Banana Strawberry".split()))
```

- This is how to find the most similar words:

```
print(model.most_similar(positive=[Prince,
Girl], negative=[Boy]))
```

A unique feature of Word2vec is that you can get vectors, from other vectors using vector operations. For example, a vector of "Prince" minus a vector of "boy" plus a vector of "girl" will be almost equal to a vector of "Princess." Hence, when you compute this, you will get a vector of "Princess."

```
Vec ("Prince") - Vec("boy") + Vec("girl") ≈
Vec("Princess")
```

This was just an example. This case is valid in many other cases. This is a specialty of Word2vec and is useful in estimating the similar words, next words, natural language generation (NLG), and so on.

Table 11-1 shows pretrained models with other parameters.

Table 11-1. Different Pretrained Models with Other Parameters

Model File	Number of Dimensions	Corpus Size	Vocabulary Size	Architecture	Context Window Size	Author
Google News	300	100B	3M	Word2Vec	BoW, ~5	Google
Freebase IDs	1000	100B	1.4M	Word2Vec, Skip-gram	BoW, ~10	Google
Freebase names	1000	100B	1.4M	Word2Vec, Skip-gram	BoW, ~10	Google
Wikipedia + Gigaword 5	50	6B	400,000	GloVe	10+10	GloVe
Wikipedia + Gigaword 5	100	6B	400,000	GloVe	10+10	GloVe
Wikipedia + Gigaword 5	200	6B	400,000	GloVe	10+10	GloVe
Wikipedia + Gigaword 5	300	6B	400,000	GloVe	10+10	GloVe
Common Crawl 42B	300	42B	1.9M	GloVe	AdaGrad	GloVe
Common Crawl 840B	300	840B	2.2M	GloVe	AdaGrad	GloVe
Wikipedia dependency	300	-	174,000	Word2Vec	Syntactic Dependencies	Levy & Goldberg
DBPedia vectors (wiki2vec)	1000	-	-	Word2Vec	BoW, 10	Idio

Building the Response

Reponses are another important part of chatbots. Based on how a chatbot replies, a user may get attracted to it. Whenever a chatbot is made, one thing that should be kept in mind is its users. You need to know who will use it and for what purpose it will be used. For example, a chatbot for a restaurant web site will be asked only about restaurants and foods. So, you know more or less what questions will be asked. Therefore, for each intent, you store multiple answers that can be used after identifying the intent so the user will not get the same answer repeatedly. You can also have one intent for any out-of-context questions; that intent can have multiple answers, and choosing randomly, the chatbot can reply.

For example, if the intent is "hello," you can have multiple replies such as "Hello! How are you?" and "Hello! How are you doing?" and "Hi! How can I help you?"

The chatbot can choose any one randomly for the reply.

In the following sample code, you are taking input from the user, but in the original chatbot, the *intent* is defined by the chatbot itself based on any question asked by the user.

```
import random
intent = input()
output = ["Hello! How are you","Hello! How are you doing","Hii!
How can I help you","Hey! There","Hiiii","Hello! How can I
assist you?","Hey! What's up?"]
if(intent == "Hii"):
 print(random.choice(output))
```

Chatbot Development Using APIs

Creating a chatbot is not an easy task. You need an eye for detail and the sharp mindedness to build a chatbot that can be put to good use. There are two approaches to building a chatbot.

- Rule-based approach

- Machine learning approach that makes the system learn on its own by streamlining data

Some chatbots are basic in nature, while others are more advanced with AI brains. Chatbots that can understand natural language and respond to them use AI brains, and technology enthusiasts are making use of various sources such as Api.ai to create such AI-rich chatbots.

Programmers are leveraging the following services to build bots:

- Microsoft bot frameworks

- Wit.ai

- Api.ai

- IBM's Watson

Other bot-building enthusiasts with limited or no programming skills are making use of bot development platforms such as the following to build chatbots:

- Chatfuel

- Texit.in

- Octane AI

- Motion.ai

There are different APIs that analyze text. The three major giants are as follows:

- Cognitive Services of Microsoft Azure

- Amazon Lex

- IBM Watson

Cognitive Services of Microsoft Azure

Let's start with Microsoft Azure.

- *Language Understanding Intelligent Service (LUIS)*: This provides simple tools that enable you to build your own language models (intents/entities) that allow any application/bot to understand your commands and act accordingly.

- *Text Analytics API*: This evaluates sentiment and topics in order to understand what users want.

- *Translator Text API*: This automatically identifies the language and then translates it into another language in real time.

- *Web Language Model API*: This inserts spaces into a string of words lacking spaces automatically.

- *Bing Spell Check API*: This enables users to correct spelling errors; recognize the difference among names, brand names, and slang; and understand homophones as they are typing.

- *Linguistic Analysis API*: This allows you to identify the concepts and actions in your text with part-of-speech tagging and find phrases and concepts using natural language parsers. It is highly useful for mining customer feedback.

Amazon Lex

Amazon Lex is a service for building conversational interfaces into any application using voice and text. Unfortunately, there is no synonym option, and there is no proper entity extraction and intent classification.

The following are some important benefits of using Amazon Lex:

- It's simple. It guides you in creating a chatbot.

- It has deep learning algorithms. Algorithms such as NLU and NLP are implemented for the chatbots. Amazon has centralized this functionality so that it can be easily used.

- It has easy deployment and scaling features.

- It has built-in integration with the AWS platform.

- It is cost effective.

IBM Watson

IBM provides the IBM Watson API to quickly build your own chatbot. In the implementation, approaching the journey is just as important as the journey itself. Educating yourself on the Watson Conversational AI for the enterprise basics of conversational design, and its impact on your business, is essential in formulating a successful plan of action. This preparation will allow you to communicate, learn, and monitor against a standard, allowing your business to build a customer-ready and successful project.

Conversational design is the most important part of building a chatbot. The first thing to understand is who the user is and what they want to achieve.

IBM Watson has many technologies that you can easily integrate in your chatbot; some of them are Watson Conversation, Watson Tone Analyzer, speech to text, and many more.

Best Practices of Chatbot Development

While building a chatbot, it is important to understand that there are certain best practices that can be leveraged. This will help in creating a successful user-friendly bot that can fulfill its purpose to have a seamless conversation with the user.

One of the foremost things in this relation is to know the target audience well. Next comes other things such as identifying the use case scenarios, setting the tone of the chat, and identifying the messaging platforms.

By adhering to the following best practices, the desire to assure seamless conversations with users can become a reality.

Know the Potential Users

A thorough understanding of the target audience is the first step in building a successful bot. The next stage is to know the purpose for which the bot is being created.

Here are some points to remember:

- Know what the purpose of the specific bot is. It could be a bot to entertain the audience, facilitate users to transact, provide news, or serve as a customer service channel.

- Make the bot more customer friendly by learning about the customer's product.

Read the User Sentiments and Make the Bot Emotionally Enriching

A chatbot should be warm and friendly just like a human in order to make the conversation a great experience. It has to smartly read as well as understand user sentiments to promote content blocks that can prompt the user to continue the conversation. The user will be encouraged to visit again if the experience is a rich one the first time.

Here are some points to remember:

- Promote your product or turn users into brand ambassadors by leveraging positive sentiments.

- Promptly address negative comments to stay afloat in the conversation game.

- Whenever possible, use friendly language to make users feel like they are interacting with a familiar human.

- Make users feel comfortable by repeating inputs and ensure that they are able to understand everything being discussed.

CHAPTER 12

Face Detection and Recognition

Face detection is the process of detecting a face in an image or video.

Face recognition is the process of detecting face in an image and then using algorithms to identify who the face belongs to. Face recognition is thus a form of person identification.

You first need to extract features from the image for training the machine learning classifier to identify faces in the image. Not only are these systems nonsubjective, but they are also *automatic*—no hand labeling of facial features is required. You simply extract features from the faces, train your classifier, and then use it to identify subsequent faces.

Since for face recognition you first need to detect a face from the image, you can think of face recognition as a two-phase stage.

- *Stage 1*: Detect the presence of faces in an image or video stream using methods such as Haar cascades, HOG + Linear SVM, deep learning, or any other algorithm that can localize faces.

- *Stage 2*: Take each of the faces detected during the localization phase and learn whom the face belongs to—this is where you actually assign a name to a face.

© Navin Kumar Manaswi 2018
N. K. Manaswi, *Deep Learning with Applications Using Python*,
https://doi.org/10.1007/978-1-4842-3516-4_12

Face Detection, Face Recognition, and Face Analysis

There is a difference between face detection, face recognition, and face analysis.

- *Face detection*: This is the technique of finding all the human faces in an image.

- *Face recognition*: This is the next step after face detection. In face recognition, you identify which face belongs to which person using an existing image repository.

- *Face analysis*: A face is examined, and some inference is taken out such as age, complexion, and so on.

OpenCV

OpenCV provides three methods for face recognition (see Figure 12-1):

- Eigenfaces

- Local binary pattern histograms (LBPHs)

- Fisherfaces

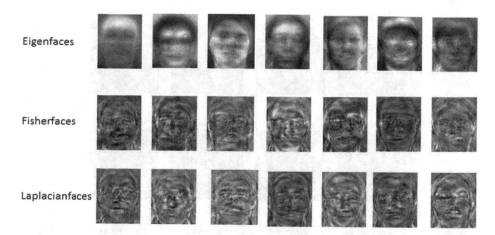

Figure 12-1. *Applying OpenCV methods to faces*

All three methods recognize a face by comparing the face with some training set of known faces. For training, you supply the algorithm with faces and label them with the person they belong to. When you use the algorithm to recognize some unknown face, it uses the model trained on the training set to make the recognition. Each of the three aforementioned methods uses the training set a bit differently.

Laplacian faces can be another way to recognize face.

Eigenfaces

The eigenfaces algorithm uses principal component analysis to construct a low-dimensional representation of face images, which you will use as features for the corresponding face images (Figure 12-2).

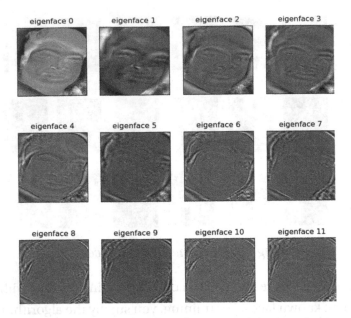

Figure 12-2. *Applying Eigenvalue decomposition and extracting 11 eigenfaces with the largest magnitude*

For this, you collect a data set of faces with multiple face images of each person you want to recognize—it's like having multiple *training examples* of an image class you want to label in image classification. With this data set of face images, presumed to be the same width and height and ideally with their eyes and facial structures aligned at the same (x, y) coordinates, you apply an eigenvalue decomposition of the data set, keeping the eigenvectors with the largest corresponding eigenvalues.

Given these eigenvectors, a face can then be represented as a linear combination of what Kirby and Sirovich called *eigenfaces*. The eigenfaces algorithm looks at the whole data set.

LBPH

You can analyze each image independently in LBPH. The LBPH method is somewhat simpler, in the sense that you characterize each image in the data set locally; when a new unknown image is provided, you perform the same analysis on it and compare the result to each of the images in the data set. The way that you analyze the images is by characterizing the local patterns in each location in the image.

While the eigenfaces algorithm relies on PCA to construct a low-dimensional representation of face images, the local binary pattern (LBP) method relies on, as the name suggests, feature extraction.

First introduced by Ahonen et al. in the 2006 paper "Face Recognition with Local Binary Patterns," the method suggests dividing a face image into a 7×7 grid of equally sized cells (Figure 12-3).

Figure 12-3. *Applying LBPH for face recognition starts by dividing the face image into a 7x7 grid of equally sized cells*

You then extract a local binary pattern histogram from each of the 49 cells. By dividing the image into cells, you introduce *locality* into the final feature vector. Furthermore, cells in the center have more weight such that they contribute *more* to the overall representation. Cells in the corners carry less identifying facial information compared to the cells in the center of the grid (which contain eyes, nose, and lip structures). Finally, you concatenate this weighted LBP histogram from the 49 cells to form your final feature vector.

Fisherfaces

The Principal Component Analysis (PCA), which is the core of the Eigenfaces method, finds a linear combination of features that maximizes the total variance in data. While this is clearly a powerful way to represent data, it doesn't consider any classes and so a lot of discriminative information may be lost when throwing components away. Imagine a situation where the variance in your data is generated by an external source, let it be the light. The components identified by a PCA do not necessarily contain any discriminative information at all, so the projected samples are smeared together and a classification becomes impossible.

The Linear Discriminant Analysis performs a class-specific dimensionality reduction and was invented by the great statistician Sir R. A. Fisher. The use of multiple measurements in taxonomic problems. In order to find the combination of features that separates best between classes the Linear Discriminant Analysis maximizes the ratio of between-classes to within-classes scatter, instead of maximizing the overall scatter. The idea is simple: same classes should cluster tightly together, while different classes are as far away as possible from each other in the lower-dimensional representation.

Detecting a Face

The first feature that you need for performing face recognition is to detect where in the current image a face is present. In Python you can use Haar cascade filters of the OpenCV library to do this efficiently.

For the implementation shown here, I used Anaconda with Python 3.5, OpenCV 3.1.0, and dlib 19.1.0. To use the following code, please make sure that you have these (or newer) versions.

To do the face detection, a couple of initializations must be done, as shown here:

```
# Import the OpenCV library
import cv2
# Initialize a face cascade using the frontal face haar cascade provided
# with the OpenCV2 library. This will be required for face detection in an
# image.
faceCascade = cv2.CascadeClassifier('haarcascade_frontalface_default.xml')
# The desired output width and height, can be modified according to the needs.
OUTPUT_SIZE_WIDTH = 700
OUTPUT_SIZE_HEIGHT = 600

# Open the first webcam device
capture = cv2.VideoCapture(0)

# Create two opencv named windows for showing the input, output images.
cv2.namedWindow("base-image", cv2.WINDOW_AUTOSIZE)
cv2.namedWindow("result-image", cv2.WINDOW_AUTOSIZE)

# Position the windows next to each other
cv2.moveWindow("base-image", 20, 200)
cv2.moveWindow("result-image", 640, 200)

# Start the window thread for the two windows we are using
cv2.startWindowThread()

rectangleColor = (0, 100, 255)
```

The rest of the code will be an infinite loop that keeps getting the latest image from the webcam, detects all faces in the image retrieved, draws a rectangle around the largest face detected, and then finally shows the input, output images in a window (Figure 12-4).

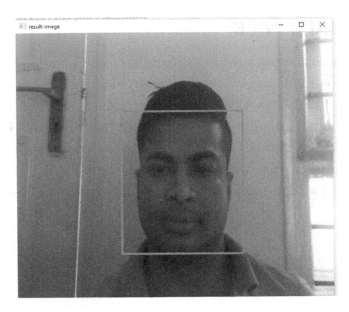

Figure 12-4. *A sample output showing detected face*

You can do this with the following code within an infinite loop:

```
# Retrieve the latest image from the webcam
rc,fullSizeBaseImage = capture.read()
# Resize the image to 520x420
baseImage= cv2.resize(fullSizeBaseImage, (520, 420))

# Check if a key was pressed and if it was Q or q, then destroy all
# opencv windows and exit the application, stopping the infinite loop.
pressedKey = cv2.waitKey(2)
if (pressedKey == ord('Q')) | (pressedKey == ord('q')):
cv2.destroyAllWindows()
exit(0)
# Result image is the image we will show the user, which is a
# combination of the original image captured from the webcam with the
# overlayed rectangle detecting the largest face
resultImage = baseImage.copy()

# We will be using gray colored image for face detection.
# So we need to convert the baseImage captured by webcam to a gray-based image
gray_image = cv2.cvtColor(baseImage, cv2.COLOR_BGR2GRAY)
# Now use the haar cascade detector to find all faces in the
# image
faces = faceCascade.detectMultiScale(gray_image, 1.3, 5)
```

```
# As we are only interested in the 'largest' face, we need to
# calculate the largest area of the found rectangle.
# For this, first initialize the required variables to 0.
maxArea = 0
x = 0
y = 0
w = 0
h = 0

# Loop over all faces found in the image and check if the area for this face is
# the largest so far
for(_x, _y, _w, _h) in faces:
if _w * _h > maxArea:
        x = _x
        y = _y
        w = _w
        h = _h
————*maxArea = w * h

# If any face is found, draw a rectangle around the
# largest face present in the picture
If maxArea > 0:
cv2.rectangle(resultImage, (x-10, y-20),
(x + w+10, y + h+20), rectangleColor, 2)
# Since we want to show something larger on the screen than the
# original 520x420, we resize the image again

# Note that it would also be possible to keep the large version
# of the baseimage and make the result image a copy of this large
# base image and use the scaling factor to draw the rectangle
# at the right coordinates.
largeResult = cv2.resize(resultImage,
(OUTPUT_SIZE_WIDTH, OUTPUT_SIZE_HEIGHT))
# Finally, we show the images on the screen
cv2.imshow("base-image", baseImage)
cv2.imshow("result-image", largeResult)
|
```

Tracking the Face

The previous code for face detection has some drawbacks.

- The code might be computationally expensive.

- If the detected person is turning their head slightly, the Haar cascade might not detect the face.

- It's difficult to keep track of a face between frames.

A better approach for this is to do the detection of the face once and then make use of the correlation tracker from the excellent dlib library to just keep track of the faces from frame to frame.

For this to work, you need to import another library and initialize additional variables.

```
import dlib

# Create the tracker we will use to recognize face in different frames
# we get from the webcam
tracker = dlib.correlation_tracker()

# The Boolean variable we use to keep track whether we are
# using dlib tracker, or not.
trackingFace = 0
```

Within the infinite for loop, you will now determine whether the dlib correlation tracker is currently tracking a region in the image. If this is *not* the case, you will use a similar code as before to find the largest face, but instead of drawing the rectangle, you use the found coordinates to initialize the correlation tracker.

```
# If we are not tracking a face, then try to detect one using the above code itself.
if not trackingFace:

# We will be using gray colored image for face detection.
# So we need to convert the baseImage captured by webcam to a gray-based image
gray = cv2.cvtColor(baseImage, cv2.COLOR_BGR2GRAY)
# Now use the haar cascade detector to find all faces
# in the image
faces=faceCascade.detectMultiScale(gray,1.3,5)

# In the console we can show our this case of using the
# detector for a face, when we are detecting it for first time.
print("Using the cascade detector to detect face")

# As we are only interested in the 'largest' face, we need to
# calculate the largest area of the found rectangle.
# For this, first initialize the required variables to 0.
maxArea = 0
x = 0
y = 0
w = 0
h =0
```

```
# Loop over all faces and check if the area for this
# face is the largest so far
# We need to convert it to int here because dlib tracker
# needs an int as its argument. If we omit the cast to
# int here, you will get cast errors since the detector
# returns numpy.int32 and the tracker requires an int
for(_x, _y, _w, _h) in faces:
if _w * _h > maxArea:
            x = int(_x)
            y = int(_y)
            w = int(_w)
            h = int(_h)
        maxArea = w * h

# If any face is found, draw a rectangle around the
# largest face present in the picture
if maxArea > 0:

#Initialize the tracker
tracker.start_track(baseImage,
dlib.rectangle(x-10, y-20, x+w+10, y+h+20))

# Set the indicator variable such that we know the
# tracker is tracking a face in the image
trackingFace = 1
```

Now the final bit within the infinite loop is to check again if the correlation tracker is actively tracking a face (i.e., did it just detect a face with the previous code, `trankingFace=1`?). If the tracker is actively tracking a face in the image, you will update the tracker. Depending on the quality of the update (i.e., how confident the tracker is about whether it is still tracking the same face), you either draw a rectangle around the region indicated by the tracker or indicate you are not tracking a face anymore.

```
# Check if the tracker is actively tracking a face in the image
if trackingFace:

# Update the tracker and request information about the
# quality of the tracking update
trackingQuality = tracker.update(baseImage)

# If the tracking quality is good enough, determine the
# updated position of the tracked region and draw the
# rectangle
If trackingQuality >= 9.0:
tracked_position = tracker.get_position()

t_x = int(tracked_position.left())
t_y = int(tracked_position.top())
t_w = int(tracked_position.width())
t_h = int(tracked_position.height())
cv2.rectangle(resultImage, (t_x,t_y),
(t_x+t_w, t_y+t_h),
rectangleColor, 2)

else:
# If the quality of the tracking update is not good enough
# for us  (e.g. the face being tracked moved out of the
# screen) we stop the tracking of the face and in the
# next loop we will find the largest face in the image
# again
trackingFace = 0
```

As you can see in the code, you print a message to the console every time you use the detector again. If you look at the output of the console while running this application, you will notice that even if you move quite a bit around on the screen, the tracker is quite good at following a face once it is detected.

Face Recognition

A face recognition system identifies the name of person present in the video frame by matching the face in each frame of video with the trained images and returns (and writes in a CSV file) the label if the face in the frame is successfully matched. You will now see how to create a face recognition system step-by-step.

First you import all the required libraries. face_recognition is the simple library built using dlib's state-of-the-art face recognition also built with deep learning.

```
import os
import re
import warnings
import scipy.misc
import cv2
import face_recognition
from PIL import Image
import argparse
import csv
import os
```

Argparse is a Python library that allows you to add your own arguments to a file; it can then be used to input any image directory or a file path at the time of execution.

```
parser = argparse.ArgumentParser()
parser.add_argument("-i", "--images-dir", help="image dir")
parser.add_argument("-v", "--video", help="video to recognize faces on")
parser.add_argument("-o", "--output-csv", help="Ouput csv file [Optional]")
parser.add_argument("-u", "--upsample-rate", help="How many times to upsample the image looking for faces. Higher numbers
                     find smaller faces. [Optional]")
args = vars(parser.parse_args())
```

In the previous code, while running this Python file, you have to specify the following: the training input image directory, video file which we will use as data set, and an output CSV file to write the output at each time frame.

```
#Check if argument values are valid
if args.get("images_dir", None) is None and os.path.exists(args.get("images_dir", None)):
    print("Please check the path to images folder")
    exit()
if args.get("video", None) is None and os.path.isfile(args.get("video", None)):
    print("Please check the path to video")
    exit()
if args.get("output_csv", None) is None:
    print("You haven't specified an output csv file. Nothing will be written.")
# By default upsample rate = 1
upsample_rate = args.get("upsample_rate", None)
if upsample_rate is None:
    upsample_rate = 1
```

```
# Helper functions
def image_files_in_folder(folder):
    return [os.path.join(folder, f) for f in os.listdir(folder) if re.match(r'.*\.(jpg|jpeg|png)', f, flags=re.I)]
```

By using the previous function, all image files from the specified folder can be read.

The following function tests the input frame with the known training images:

```
def test_image(image_to_check, known_names, known_face_encodings, number_of_times_to_upsample=1):
    """
    Test if any face is recognized in unknown image by checking known images
    :paramimage_to_check: Numpy array of the image
    :paramknown_names: List containing known labels
    :paramknown_face_encodings: List containing training image labels
    :paramnumber_of_times_to_upsample: How many times to upsample the image looking for
    faces. Higher numbers find smaller faces.
    :return: A list of labels of known names
    """
    # unknown_image = face_recognition.load_image_file(image_to_check)
    unknown_image = image_to_check
    # Scale down the image to make it run faster
    if unknown_image.shape[1] > 1600:
        scale_factor = 1600 / unknown_image.shape[1]
        with warnings.catch_warnings():
            warnings.simplefilter("ignore")
            unknown_image = scipy.misc.imresize(unknown_image, scale_factor)
    face_locations = face_recognition.face_locations(unknown_image, number_of_times_to_upsample)
    unknown_encodings = face_recognition.face_encodings(unknown_image, face_locations)
    result = []
    for unknown_encoding in unknown_encodings:
        result = face_recognition.compare_faces(known_face_encodings, unknown_encoding)
    result_encoding = []
    for nameIndex, is_match in enumerate(result):
        if is_match:
            result_encoding.append(known_names[nameIndex])

    return result_encoding
```

Now you define the function to extract the label for matched, known images.

```
def map_file_pattern_to_label(labels_with_pattern, labels_list):
    """
    Map file name pattern to full label
    :paramlabels_with_pattern: dict : { "file_name_pattern": "full_label" }
    :paramlabels_list: list : list of labels of file names got from test_image()
    :return: list of full labels
    """
    result_list = []
    for key, label in labels_with_pattern.items():
        for img_labels in labels_list:
            if str(key).lower() in str(img_labels).lower():
                if str(label) not in result_list:
                    result_list.append(str(label))
                # continue
    # result_list = [label for key, label in labels_with_pattern if str(key).lover() in labels_list]
    return result_list
```

Read the input video to extract test frames.

```
cap = cv2.VideoCapture(args["video"])

#get the training images
training_encodings = []
training_labels = []
for file in image_files_in_folder(args['images_dir']):
    basename = os.path.splitext(os.path.basename(file))[0]
    img = face_recognition.load_image_file(file)
    encodings = face_recognition.face_encodings(img)

    if len(encodings) > 1:
        print("WARNING: More than one face found in {}. Only considering the first face.".format(file))

    if len(encodings) == 0:
        print("WARNING: No faces found in {}. Ignoring file.".format(file))
    if len(encodings):
        training_labels.append(basename
        training_encodings.append(encodings[0])

csvfile = None
csvwriter = None
if args.get("output_csv", None) is not None:
    csvfile = open(args.get("output_csv"), 'w')
    csvwriter = csv.writer(csvfile, delimiter=',', quotechar='|', quoting=csv.QUOTE_MINIMAL)

ret, firstFrame = cap.read()
frameRate = cap.get(cv2.CAP_PROP_FPS)
```

Now define the labels of your training sets. Then match the extracted frame from the given input video to get the desired results.

```
# Labels with file pattern, edit this
label_pattern = {
    "shah": "Shahrukh Khan",
    "amir": "Amir Khan"
            }

# match each frame in video with our trained set of labeled images
while ret:
    curr_frame = cap.get(1)

    ret, frame = cap.read()

    result = test_image(frame, training_labels, training_encodings, upsample_rate)
    labels = map_file_pattern_to_label(label_pattern, result)
    curr_time = curr_frame / frameRate
    print("Time: {} faces: {}".format(curr_time, labels))
    if csvwriter:
        csvwriter.writerow([curr_time, labels])
    cv2.imshow('frame', frame)

    key = cv2.waitKey(1) & 0xFF
    if key == ord('q'):
        break
if csvfile:
    csvfile.close()
cap.release()
cv2.destroyAllWindows()
```

Deep Learning–Based Face Recognition

Import the necessary packages.

```
import cv2                  # working with, mainly resizing, images
import numpy as np          # dealing with arrays
import os                   # dealing with directories
from random import shuffle  # mixing up or currently ordered data that might lead our network astray in training.
from tqdm import tqdm
from scipy import misc
import tflearn
from tflearn.layers.conv import conv_2d, max_pool_2d
from tflearn.layers.core import input_data, dropout, fully_connected
from tflearn.layers.estimator import regression
import tensorflow as tf
import glob
import matplotlib.pyplot as plt
import dlib
```

Initialize the variables.

```
from skimage import io
tf.reset_default_graph()
TRAIN_DIR ='resize_a/train'
TEST_DIR ='resize_a/test'
IMG_SIZE = 200
boxScale=1
LR = 1e-3
MODEL_NAME = 'quickest.model'.format(LR, '2conv-basic')
```

The `label_img()` function is used to create the label array, and the `detect_faces()` function detects the face portion in the image.

```
def label_img(img):
    word = img.split('(')[-2]
    word_label = word[0]
    if word_label == 'R': return [1,0]

    elif word_label == 'A': return [0,1]

def detect_faces(image):

    # Create a face detector
    face_detector = dlib.get_frontal_face_detector()

    # Run detector and get bounding boxes of the faces on image.
    detected_faces = face_detector(image, 1)
    face_frames = [(x.left(), x.top(),
                    x.right(), x.bottom()) for x in detected_faces]

    return face_frames
```

The `create_train_data()` function is used for preprocessing the training data.

```
def create_train_data():
    training_data = []
    for img in tqdm(os.listdir(TRAIN_DIR)):
        label = label_img(img)
        path = os.path.join(TRAIN_DIR,img)
        img= misc.imread(path)
        img = cv2.imread(path,cv2.IMREAD_GRAYSCALE)
        img = cv2.resize(img, (IMG_SIZE,IMG_SIZE))
        detected_faces = detect_faces(img)
        for n, face_rect in enumerate(detected_faces):
            img = Image.fromarray(img).crop(face_rect)
            img = np.array(img)
        img = cv2.resize(img, (IMG_SIZE,IMG_SIZE))
# If any face is found, draw a rectangle around the
#  largest face present in the picture

        training_data.append([np.array(img),np.array(label)])
    shuffle(training_data)
    np.save('train_data.npy', training_data)
    return training_data
```

The `process_test_data()` function is used to preprocess the testing data.

```
def process_test_data():
    testing_data = []
    for img in tqdm(os.listdir(TEST_DIR)):
        path = os.path.join(TEST_DIR,img)
        imgnum = img.split('.')[-2]
        img_num=get_num(imgnum)
        img= misc.imread(path)
        img = cv2.imread(path,cv2.IMREAD_GRAYSCALE)
        img = cv2.resize(img, (IMG_SIZE,IMG_SIZE))
        detected_faces = detect_faces(img)
        for n, face_rect in enumerate(detected_faces):
            img = Image.fromarray(img).crop(face_rect)
            img = np.array(img)
        img = cv2.resize(img, (IMG_SIZE,IMG_SIZE))
# If any face is found, draw a rectangle around the
# largest face present in the picture

        testing_data.append([np.array(img), img_num])
```

Then you create the model and fit the training data in the model.

```
train_data= create_train_data()
train = train_data[:-2]
test = train_data[-2:]
X = np.array([i[0] for i in train]).reshape(-1,200,200,1)
Y = [i[1] for i in train]
test_x = np.array([i[0] for i in test]).reshape(-1,200,200,1)
test_y = [i[1] for i in test]
convnet = input_data(shape=[None, 200, 200, 1], name='input')

convnet = conv_2d(convnet, 4, 5, activation='relu')
convnet = max_pool_2d(convnet, 5)

convnet = conv_2d(convnet, 5, 5, activation='relu')
convnet = max_pool_2d(convnet, 5)

convnet = conv_2d(convnet, 8, 5, activation='relu')
convnet = max_pool_2d(convnet, 5)

convnet = fully_connected(convnet, 8, activation='relu')
convnet = dropout(convnet, 0.2)

convnet = fully_connected(convnet, 2, activation='softmax')
convnet = regression(convnet, optimizer='adam', learning_rate=LR, loss='categorical_crossentropy', name='targets')
model.fit({'input': X}, {'targets': Y}, n_epoch=1, validation_set=({'input': test_x}, {'targets': test_y}),
    snapshot_step=500, show_metric=True, run_id=MODEL_NAME)
```

Finally, you prepare the test data and predict the output.

```
test_data = process_test_data()

fig=plt.figure()

for num,data in enumerate(test_data[:12]):

    img_num = data[1]
    img_data = data[0]
    y = fig.add_subplot(3,4,num+1)
    orig = img_data
    data = img_data.reshape(IMG_SIZE,IMG_SIZE,1)
    #model_out = model.predict([data])[0]
    model_out = model.predict([data])[0]

    if np.argmax(model_out) == 0: str_label='Ronaldo'
    elif np.argmax(model_out) == 1: str_label='amitabh'

    y.imshow(orig,cmap='gray')
    plt.title(str_label)
    y.axes.get_xaxis().set_visible(False)
    y.axes.get_yaxis().set_visible(False)
plt.show()
```

Transfer Learning

Transfer learning makes use of the knowledge gained while solving one problem and applying it to a different but related problem.

Here you will see how you can use a pretrained deep neural network called the Inception v3 model for classifying images.

The Inception model is quite capable of extracting useful information from an image.

Why Transfer Learning?

It's well known that convolutional networks require significant amounts of data and resources to train.

It has become the norm for researchers and practitioners alike to use transfer learning and fine-tuning (that is, transferring the network weights trained on a previous project such as ImageNet to a new task).

You can take two approaches.

- *Transfer learning*: You can take a CNN that has been pretrained on ImageNet, remove the last fully connected layer, and then treat the rest of the CNN as a feature extractor for the new data set. Once you extract the features for all images, you train a classifier for the new data set.

- *Fine-tuning*: You can replace and retrain the classifier on top of the CNN and also fine-tune the weights of the pretrained network via backpropagation.

Transfer Learning Example

In this example, first you will try to classify images by directly loading the Inception v3 model.

Import all the required libraries.

```
%matplotlib inline
import matplotlib.pyplot as plt
import tensorflow as tf
import numpy as np
import os

# Functions and classes for loading and using the Inception model.
import inception
```

Now define the storage directory for the model and then download the Inception v3 model.

```
inception.data_dir = 'D:/'
```

```
inception.maybe_download()
```

Load the pretrained model and define the function to classify any given image.

```
model = inception.Inception()


def classify(image_path):
    # Display the image.
    p = Image.open(image_path)
    p.show()

    # Use the Inception model to classify the image.
    pred = model.classify(image_path=image_path)

    # Print the scores and names for the top-10 predictions.
    model.print_scores(pred=pred, k=10, only_first_name=True)
```

Now that the model is defined, let's check it for some images.

This gives a 91.11 percent correct result, but now if you check for some person, this is what you get:

It's 48.50 percent tennis ball!

Unfortunately, the Inception model seemed unable to classify images of people. The reason for this was the data set used for training the Inception model, which had some confusing text labels for classes.

You can instead reuse the pretrained Inception model and merely replace the layer that does the final classification. This is called *transfer learning*.

First you input and process an image with the Inception model. Just prior to the final classification layer of the Inception model, you save the so-called transfer values to a cache file.

The reason for using a cache file is that it takes a long time to process an image with the Inception model. When all the images in the new data set have been processed through the Inception model and the resulting transfer values are saved to a cache file, then you can use those transfer values as the input to another neural network. You will then train the second neural network using the classes from the new data set, so the network learns how to classify images based on the transfer values from the Inception model.

In this way, the Inception model is used to extract useful information from the images, and another neural network is then used for the actual classification.

Calculate the Transfer Value

Import the `transfer_value_cache` function from the Inception file.

```
from inception import transfer_values_cache
```

```
file_path_cache_train = os.path.join(cifar10.data_path, 'inception_cifar10_train.pkl')
file_path_cache_test = os.path.join(cifar10.data_path, 'inception_cifar10_test.pkl')
```

```
print("Processing Inception transfer-values for training-images ...")

# Scale images because Inception needs pixels to be between 0 and 255,
# while the CIFAR-10 functions return pixels between 0.0 and 1.0
images_scaled = images_train * 255.0

# If transfer-values have already been calculated then reload them,
# otherwise calculate them and save them to a cache-file.
transfer_values_train = transfer_values_cache(cache_path=file_path_cache_train,
                                              images=images_scaled,
                                              model=model)
```

```
Processing Inception transfer-values for training-images ...
- Processing image:   1021 / 50000
```

```
print("Processing Inception transfer-values for test-images ...")

# Scale images because Inception needs pixels to be between 0 and 255,
# while the CIFAR-10 functions return pixels between 0.0 and 1.0
images_scaled = images_test * 255.0

# If transfer-values have already been calculated then reload them,
# otherwise calculate them and save them to a cache-file.
transfer_values_test = transfer_values_cache(cache_path=file_path_cache_test,
                                             images=images_scaled,
                                             model=model)
```

As of now, the transfer values are stored in the cache file. Now you will create a new neural network.

Define the networks.

```
# Wrap the transfer-values as a Pretty Tensor object.
x_pretty = pt.wrap(x)

with pt.defaults_scope(activation_fn=tf.nn.relu):
    y_pred, loss = x_pretty.\
          fully_connected(size=1024, name='layer_fc1').\
          softmax_classifier(num_classes=num_classes, labels=y_true)
```

Here is the optimization method:

```
global_step = tf.Variable(initial_value=0,
                          name='global_step', trainable=False)
optimizer = tf.train.AdamOptimizer(learning_rate=1e-4).minimize(loss, global_step)
```

Here is the classification accuracy:

```
y_pred_cls = tf.argmax(y_pred, dimension=1)
correct_prediction = tf.equal(y_pred_cls, y_true_cls)
accuracy = tf.reduce_mean(tf.cast(correct_prediction, tf.float32))
```

Here is the TensorFlow run:

```
session = tf.Session()
session.run(tf.global_variables_initializer())
```

Here is the helper function to perform batch training:

```
def random_batch():
    # Number of images (transfer-values) in the training-set.
    num_images = len(transfer_values_train)

    # Create a random index.
    idx = np.random.choice(num_images,
                           size=train_batch_size,
                           replace=False)

    # Use the random index to select random x and y-values.
    # We use the transfer-values instead of images as x-values.
    x_batch = transfer_values_train[idx]
    y_batch = labels_train[idx]

    return x_batch, y_batch
```

For optimizing, here is the code:

```
def optimize(num_iterations):
    # Number of images (transfer-values) in the training-set.

    start_time = time.time()

    for i in range(num_iterations):
        # Get a batch of training examples.
        # x_batch now holds a batch of images (transfer-values) and
        # y_true_batch are the true labels for those images.
        x_batch, y_true_batch = random_batch()

        # Put the batch into a dict with the proper names
        # for placeholder variables in the TensorFlow graph.
        feed_dict_train = {x: x_batch,
                           y_true: y_true_batch}

        # Run the optimizer using this batch of training data.
        # TensorFlow assigns the variables in feed_dict_train
        # to the placeholder variables and then runs the optimizer.
        # We also want to retrieve the global_step counter.
        i_global, _ = session.run([global_step, optimizer],
                                  feed_dict=feed_dict_train)

        # Print status to screen every 100 iterations (and last).
        if (i_global % 100 == 0) or (i == num_iterations - 1):
            # Calculate the accuracy on the training-batch.
            batch_acc = session.run(accuracy,
                                    feed_dict=feed_dict_train)
```

```
            # Print status.
            msg = "Global Step: {0:>6}, Training Batch Accuracy: {1:>6.1%}"
            print(msg.format(i_global, batch_acc))

    # Ending time.
    end_time = time.time()

    # Difference between start and end-times.
    time_dif = end_time - start_time

    # Print the time-usage.
    print("Time usage: " + str(timedelta(seconds=int(round(time_dif)))))

    # Use the random index to select random x and y-values.
    # We use the transfer-values instead of images as x-values.
    x_batch = transfer_values_train[idx]
    y_batch = labels_train[idx]

    return x_batch, y_batch
```

For plotting the confusion matrix, here is the code:

```
from sklearn.metrics import confusion_matrix

def plot_confusion_matrix(cls_pred):
    # This is called from print_test_accuracy() below.

    # cls_pred is an array of the predicted class-number for
    # all images in the test-set.

    # Get the confusion matrix using sklearn.
    cm = confusion_matrix(y_true=cls_test,    # True class for test-set.
                          y_pred=cls_pred)    # Predicted class.

    # Print the confusion matrix as text.
    for i in range(num_classes):
        # Append the class-name to each line.
        class_name = "({}) {}".format(i, class_names[i])
        print(cm[i, :], class_name)

    # Print the class-numbers for easy reference.
    class_numbers = [" ({0})".format(i) for i in range(num_classes)]
    print("".join(class_numbers))
```

Here is the helper function for calculating the classifications:

```python
# Split the data-set in batches of this size to limit RAM usage.
batch_size = 256

def predict_cls(transfer_values, labels, cls_true):
    # Number of images.
    num_images = len(transfer_values)

    # Allocate an array for the predicted classes which
    # will be calculated in batches and filled into this array.
    cls_pred = np.zeros(shape=num_images, dtype=np.int)

    # Now calculate the predicted classes for the batches.
    # We will just iterate through all the batches.
    # There might be a more clever and Pythonic way of doing this.

    # The starting index for the next batch is denoted i.
    i = 0

    while i < num_images:
        # The ending index for the next batch is denoted j.
        j = min(i + batch_size, num_images)

        # Create a feed-dict with the images and labels
        # between index i and j.
        feed_dict = {x: transfer_values[i:j],
                     y_true: labels[i:j]}

        # Calculate the predicted class using TensorFlow.
        cls_pred[i:j] = session.run(y_pred_cls, feed_dict=feed_dict)

        # Set the start-index for the next batch to the
        # end-index of the current batch.
        i = j

    # Create a boolean array whether each image is correctly classified.
    correct = (cls_true == cls_pred)

    return correct, cls_pred
```

```python
def classification_accuracy(correct):
    # When averaging a boolean array, False means 0 and True means 1.
    # So we are calculating: number of True / len(correct) which is
    # the same as the classification accuracy.

    # Return the classification accuracy
    # and the number of correct classifications.
    return correct.mean(), correct.sum()
```

```python
def predict_cls_test():
    return predict_cls(transfer_values = transfer_values_test,
                       labels = labels_test,
                       cls_true = cls_test)
```

```
def print_test_accuracy(show_example_errors=False,
                        show_confusion_matrix=False):

    # For all the images in the test-set,
    # calculate the predicted classes and whether they are correct.
    correct, cls_pred = predict_cls_test()

    # Classification accuracy and the number of correct classifications.
    acc, num_correct = classification_accuracy(correct)

    # Number of images being classified.
    num_images = len(correct)

    # Print the accuracy.
    msg = "Accuracy on Test-Set: {0:.1%} ({1} / {2})"
    print(msg.format(acc, num_correct, num_images))

    # Plot some examples of mis-classifications, if desired.
    if show_example_errors:
        print("Example errors:")
        plot_example_errors(cls_pred=cls_pred, correct=correct)

    # Plot the confusion matrix, if desired.
    if show_confusion_matrix:
        print("Confusion Matrix:")
        plot_confusion_matrix(cls_pred=cls_pred)
```

Now let's run it.

```
from datetime import timedelta

optimize(num_iterations=1000)

Global Step:  13100, Training Batch Accuracy: 100.0%
Global Step:  13200, Training Batch Accuracy: 100.0%
Global Step:  13300, Training Batch Accuracy: 100.0%
Global Step:  13400, Training Batch Accuracy: 100.0%
Global Step:  13500, Training Batch Accuracy: 100.0%
Global Step:  13600, Training Batch Accuracy: 100.0%
Global Step:  13700, Training Batch Accuracy: 100.0%
Global Step:  13800, Training Batch Accuracy: 100.0%
Global Step:  13900, Training Batch Accuracy: 100.0%
Global Step:  14000, Training Batch Accuracy: 100.0%
Time usage: 0:00:36

print_test_accuracy(show_example_errors=True,
show_confusion_matrix=True)
```

```
Accuracy on Test-Set: 83.2% (277 / 333)
Example errors:
Confusion Matrix:
[108 3 5] (0) Aamir Khan
[0 83 22] (1) Salman Khan
[4 22 86] (2) Shahrukh Khan
 (0) (1) (2)
```

APIs

Many easy-to-use APIs are also available for the tasks of face detection and face recognition.

Here are some examples of face detection APIs:

- PixLab

- Trueface.ai

- Kairos

- Microsoft Computer Vision

Here are some examples of face recognition APIs:

- Face++

- LambdaLabs

- KeyLemon

- PixLab

If you want face detection, face recognition, and face analysis from one provider, currently there are three major giants that are leading here.

- Amazon's Amazon Recognition API

- Microsoft Azure's Face API

- IBM Watson's Visual Recognition API

Amazon's Amazon Recognition API can do four types of recognition.

- *Object and scene detection*: Recognition identifies various interesting objects such as vehicles, pets, or furniture, and it provides a confidence score.

- *Facial analysis*: You can locate faces within images and analyze face attributes, such as whether the face is smiling or the eyes are open, with certain confidence scores.

- *Face comparison*: Amazon's Amazon Recognition API lets you measure the likelihood that faces in two images are of the same person. Unfortunately, the similarity measure of two faces of the same person depends on the age at the time of the photos. Also, a localized increase in the illumination of a face alters the results of the face comparison.

- *Facial recognition*: The API identifies the person in a given image using a private repository. It is fast and accurate.

Microsoft Azure's Face API will return a confidence score for how likely it is that the two faces belong to one person. Microsoft also has other APIs such as the following:

- *Computer Vision API*: This feature returns information about visual content found in an image. It can use tagging, descriptions, and domain-specific models to identify content and label it with confidence.

- *Content Moderation API*: This detects potentially offensive or unwanted images, text in various languages, and video content.

- *Emotion API*: This analyzes faces to detect a range of feelings and personalize your app's responses.

- *Video API*: This produces stable video output, detects motion, creates intelligent thumbnails, and detects and tracks faces.

- *Video Indexer*: This finds insights in video such as entities of speech, sentiment polarity of speech, and audio timeline.

- *Custom Vision Service*: This tags a new image based on the built-in models or the models built through training data sets provided by you.

IBM Watson's Visual Recognition API can do some specific detection such as the following:

- It can determine the age of the person.

- It can determine the gender of the person.

- It can determine the location of the bounding box around a face.

- It can return information about a celebrity who is detected in the image. (This is not returned when a celebrity is not detected.)

APPENDIX 1

Keras Functions for Image Processing

Keras has a function called ImageDataGenerator that provides you with batches of tensor image data with real-time data augmentation. Data will be looped over in batches indefinitely.

Here is the function:

```
keras.preprocessing.image.ImageDataGenerator
    (featurewise_center=False,
     samplewise_center=False,
     featurewise_std_normalization=False,
     samplewise_std_normalization=False,
     zca_whitening=False,
     zca_epsilon=1e-6,
     rotation_range=0.,
     width_shift_range=0.,
     height_shift_range=0.,
     shear_range=0.,
     zoom_range=0.,
     channel_shift_range=0.,
     fill_mode='nearest',
     cval=0.,
     horizontal_flip=False,
     vertical_flip=False,
     rescale=None,
     preprocessing_function=None,
     data_format=K.image_data_format())
```

© Navin Kumar Manaswi 2018
N. K. Manaswi, *Deep Learning with Applications Using Python*,
https://doi.org/10.1007/978-1-4842-3516-4

Here are the function's arguments:

- `featurewise_center`: Data type `boolean`. Sets input mean to 0 over the data set, feature-wise.

- `samplewise_center`: Data type `boolean`. Sets each sample mean to 0.

- `featurewise_std_normalization`: Data type `boolean`. Divides inputs by `std` of the data set, feature-wise.

- `samplewise_std_normalization`: Data type `boolean`. Divides each input by its `std`.

- `zca_epsilon`: Epsilon for ZCA whitening. The default is `1e-6`.

- `zca_whitening`: `boolean`. Applies ZCA whitening.

- `rotation_range`: `int`. Sets degree of range for random rotations.

- `width_shift_range`: Data type `float` (fraction of total width). Sets range for random horizontal shifts.

- `height_shift_range`: Data type `float` (fraction of total height). Sets range for random vertical shifts.

- `shear_range`: Data type `float`. Sets shear intensity (shear angle in counterclockwise direction as radians).

- `zoom_range`: Data type `float` or `[lower, upper]`. Sets range for random zoom. If a float, `[lower, upper]` = `[1-zoom_range, 1+zoom_range]`.

- `channel_shift_range`: Data type `float`. Sets range for random channel shifts.

- `fill_mode`: One of {`"constant"`, `"nearest"`, `"reflect"` or `"wrap"`}. Points outside the boundaries of the input are filled according to the given mode.

- `cval`: Data type `float` or `int`. The value is used for points outside the boundaries when `fill_mode` = `"constant"`.

- `horizontal_flip`: Data type `boolean`. Randomly flips inputs horizontally.

- `vertical_flip`: Data type `boolean`. Randomly flips inputs vertically.

- `rescale`: Rescaling factor. This defaults to `None`. If `None` or 0, no rescaling is applied. Otherwise, you multiply the data by the value provided (before applying any other transformation).

- `preprocessing_function`: Function that will be implied on each input. The function will run before any other modification on it. The function should take one argument, an image (a Numpy tensor with the rank 3), and should output a Numpy tensor with the same shape.

- `data_format`: One of {`"channels_first"`, `"channels_last"`}. `"channels_last"` mode means that the images should have shape (`samples`, `height`, `width`, `channels`). `"channels_first"` mode means that the images should have shape (`samples`, `channels`, `height`, `width`). It defaults to the `image_data_format` value found in your Keras config file at `~/.keras/keras.json`. If you do not set it, then it will be `"channels_last"`.

Here are its methods:

- `fit(x)`: Computes the internal data stats related to the data-dependent transformations, based on an array of sample data. This is required only if it's `featurewise_center` or `featurewise_std_normalization` or `zca_whitening`.

  - Here are the method's arguments:

    - x: Sample data. This should have a rank of 4. In the case of grayscale data, the channel's axis should have a value of 1, and in the case of RGB data, it should have a value of 3.

    - augment: Data type `boolean` (default: `False`). This sets whether to fit on randomly augmented samples.

    - rounds: Data type `int` (default: 1). If `augment` is set, this sets how many augmentation passes over the data to use.

    - seed: Data type `int` (default: `None`). Sets a random seed.

- `flow(x, y)`: Takes Numpy data and label arrays and generates batches of augmented/normalized data. Yields batches indefinitely, in an infinite loop.

  - Here are its arguments:

    - x: Data. This should have the rank 4. In the case of grayscale data, the channel's axis should have a value of 1, and in the case of RGB data, it should have a value of 3.

    - y: Labels.

- `batch_size`: Data type `int` (default: 32).

- `shuffle`: Data type `boolean` (default: `True`).

- `seed`: Data type `int` (default: `None`).

- `save_to_dir`: `None` or `str` (default: `None`). This allows you to optimally specify a directory to which to save the augmented pictures being generated (useful for visualizing what you are doing).

- `save_prefix`: Data type `str` (default: `' '`). This is the prefix to use for file names of saved pictures (relevant only if `save_to_dir` is set).

- `save_format`: Either `png` or `jpeg` (relevant only if `save_to_dir` is set). Default: `png`.

- `yields`: Tuples of (`x, y`) where `x` is a Numpy array of image data and `y` is a Numpy array of corresponding labels. The generator loops indefinitely.

The function will help you augment image data in real time, during the training itself, by creating batches of images. This will be passed during the training time.

The processing function can be used to write some manual functions also, which are not provided in the Keras library.

APPENDIX 2

Some of the Top Image Data Sets Available

- **MNIST**: Perhaps the most famous image data set available to you, this data set was compiled by Yann LeCun and team. This data set is used almost everywhere as a tutorial or introduction in computer vision. It has some 60,000 training images and about 10,000 test images.

- **CIFAR-10**: This data set was made extremely famous by the ImageNet challenge. It has 60,000 32×32 images in 10 classes, with 6,000 images per class. There are 50,000 training images and 10,000 test images.

- **ImageNet**: This labeled object image database is used in the ImageNet Large Scale Visual Recognition Challenge. It includes labeled objects, bounding boxes, descriptive words, and SIFT features. There are a total of 14,197,122 instances.

- **MS COCO**: The Microsoft Common Objects in COntext (MS COCO) data set contains 91 common object categories, with 82 of them having more than 5,000 labeled instances. In total, the data set has 2,500,000

© Navin Kumar Manaswi 2018
N. K. Manaswi, *Deep Learning with Applications Using Python*,
https://doi.org/10.1007/978-1-4842-3516-4

labeled instances in 328,000 images. In contrast to the popular ImageNet data set, COCO has fewer categories but more instances per category. COCO is a large-scale object detection, segmentation, and captioning data set.

- **10k US Adult Faces**: This data set contains 10,168 natural phace photographs and several measures for 2,222 of the faces, including memorability scores, computer vision and physical attributes, and landmark point annotations.

- **Flickr 32/47 Brands Logos**: This consists of real-world images collected from Flickr of company logos in various circumstances. It comes in two versions: the 32-brand data set and the 47-brand data set. There are a total of 8,240 images.

- **YouTube Faces**: This is a database of face videos designed for studying the problem of unconstrained face recognition in videos. The data set contains 3,425 videos of 1,595 different people.

- **Caltech Pedestrian**: The Caltech Pedestrian data set consists of approximately 10 hours of 640×480 30Hz video taken from a vehicle driving through regular traffic in an urban environment. About 250,000 frames (in 137 approximately minute-long segments) with a total of 350,000 bounding boxes and 2,300 unique pedestrians were annotated.

- **PASCAL VOC**: This is a huge data set for the image classification task. It has 500,000 instances of data.

- **Microsoft Common Objects in Context (COCO):** It contains complex everyday scenes of common objects in their natural context. Object highlighting, labeling, and classification into 91 object types. It contains 2,500,000 instances.

- **Caltech-256**: This is a large data set of images for object classification. Images are categorized and hand-sorted. There are a total of 30,607 images.

- **FBI crime data set**: The FBI crime data set is amazing. If you are interested in time-series data analysis, you can use it to plot changes in crime rates at the national level over a 20-year period.

Medical Imaging: DICOM File Format

Digital Imaging and Communication in Medicine (DICOM) is a type of file format used in the medical domain to store or transfer images taken during various tests of multiple patients.

Why DICOM?

MRIs, CT scans, and X-rays can be stored in a normal file format, but because of the uniqueness of a medical report, many different types of data are required for a particular image.

What Is the DICOM File Format?

This file format contains a header consisting of metadata of the image such as the patient's name, ID, blood group, and so on. It also contains space-separated pixel values of the images taken during various medical tests.

© Navin Kumar Manaswi 2018
N. K. Manaswi, *Deep Learning with Applications Using Python*,
https://doi.org/10.1007/978-1-4842-3516-4

The DICOM standard is a complex file format that can be handled by the following packages:

- pydicom: This is a package for working with images in Python. dicom was the older version of this package. As of this writing, pydicom 1.*x* is the latest version.

- oro.dicom: This is a package for working with images in R.

DICOM files are represented as FileName.dcm

```
import dicom

ds = dicom.read_file("E:/datasciencebowl/stage1/00cba091fa4ad62cc3200a657aeb957e/0a291d1b12b86213d813e3796f14b329.dcm")
```

```
(0008, 0005) Specific Character Set          CS: 'ISO_IR 100'
(0008, 0016) SOP Class UID                   UI: CT Image Storage
(0008, 0018) SOP Instance UID                UI: 1.2.840.113654.2.55.158283083714550104456272463610634335 9
(0008, 0060) Modality                        CS: 'CT'
(0008, 103e) Series Description              LO: 'Axial'
(0010, 0010) Patient's Name                  PN: '00cba091fa4ad62cc3200a657aeb957e'
(0010, 0020) Patient ID                      LO: '00cba091fa4ad62cc3200a657aeb957e'
(0010, 0030) Patient's Birth Date            DA: '19000101'
(0018, 0060) KVP                             DS: ''
(0020, 000d) Study Instance UID              UI: 2.25.86206730140539712382771890501772734277950692397709007305473
(0020, 000e) Series Instance UID             UI: 2.25.115758773296352289258085968002699747408935194517846 26046614
(0020, 0011) Series Number                   IS: '3'
(0020, 0012) Acquisition Number              IS: '1'
(0020, 0013) Instance Number                 IS: '88'
(0020, 0020) Patient Orientation             CS: ''
(0020, 0032) Image Position (Patient)        DS: ['-145.500000', '-158.199997', '-241.199997']
(0020, 0037) Image Orientation (Patient)     DS: ['1.000000', '0.000000', '0.000000', '0.000000', '1.000000', '0.000(
']
(0020, 0052) Frame of Reference UID          UI: 2.25.83033509634441686385652073462983801840121916678417719669650
(0020, 1040) Position Reference Indicator    LO: 'SN'
(0020, 1041) Slice Location                  DS: '-241.199997'
(0028, 0002) Samples per Pixel               US: 1
(0028, 0004) Photometric Interpretation      CS: 'MONOCHROME2'
(0028, 0010) Rows                            US: 512
(0028, 0011) Columns                         US: 512
(0028, 0030) Pixel Spacing                   DS: ['0.597656', '0.597656']
(0028, 0100) Bits Allocated                  US: 16
```

Index

© Navin Kumar Manaswi 2018
N. K. Manaswi, *Deep Learning with Applications Using Python*,
https://doi.org/10.1007/978-1-4842-3516-4

Get the eBook for only $5!

Why limit yourself?

With most of our titles available in both PDF and ePUB format, you can access your content wherever and however you wish—on your PC, phone, tablet, or reader.

Since you've purchased this print book, we are happy to offer you the eBook for just $5.

To learn more, go to http://www.apress.com/companion or contact support@apress.com.

Apress®

Printed in the United States
By Bookmasters